FOR THE YOUTH OF TODAY,

"For the youth of today, may you find true purpose throughout each challenge and infinite strength in the vulnerabilities that arise. Knowing it is through the open doorway of our heart that we are able to create a brand-new world that equally supports and inspires all".

—**MATT KAHN**
Best-selling author of *Whatever Arises, Love That* and *Everything is Here to Help You*; Spiritual teacher, empathic healer and YouTube sensation, U.S.A.

"Water through every aspect of our lives, following that flow, makes us realize that everything is so vitally interconnected. Let the horizon be the beginning of your world, not your destination; embark on the journey and follow your heart, it knows the way, better than any other voice you might encounter along the way trying to change you because of their fear not because of your risk".

—**CRISTINA ZENATO**
Shark Whisperer, Ocean Conservationist and founder of *People of the Water*, The Bahamas

"As young people you will now have to guide us through the next generation. It will be up to you to take care of Mother Earth, to heal her, to help her flourish. Our way of life is living in harmony with the winged ones, the swimmers, the fliers, the crawlers, the Root nation, the Plant nation and the Tree nation and our Grandfather Stones, the four-legged and the two-legged for we are all one, and we must come back together as one, to walk the Sacred way".

—**ELDER WHABAGOON** (Flower Blooming in Spring)
Ojibway Elder with the Loon Clan and member of the Lac Seul First Nation, Canada

"Si el mundo trabajara conexión con mundo espiritual, la vida cambia en el mundo material, y si cambia, comenzaremos actuar mas despierto para bien de la humanidad".

—**MANARI USHIGUA**
High Shaman, spiritual teacher, healer and leader of the Sápara Nation, Amazon, Ecuador

"A holistic approach and interaction with our environment will always produce a tight bond, deeper understanding and respect, and create a better place for all. I am so glad that you have dedicated this book to youth, our future".

—**JACQUES NDOUTOUMVÉ**
Vice President of the Canada-Africa Chamber of Business, Canada-Gabon

A New HUMAN STORY

A Co-Creator's Guide to Living Our True Potential

SONIA MOLODECKY

Foreword by Arzu Mountain Spirit
Blessings by Elder Whabagoon

CONDORIA

A New Human Story: A Co-Creator's Guide to Living Our True Potential

Copyright 2021 by Sonia Molodecky

ISBN: 978-1-7774672-0-3

All rights reserved.

Book Designer: DigiWriting

Cover Image and Illustrations: Iryna Molodecky

Published in Stratford, Canada

Printed and bound in Canada.

The author greatly appreciates you taking the time to read this work. Please consider leaving a review wherever you bought the book, or telling your friends or blog readers about *A New Human Story: A Co-Creator's Guide to Living Our True Potential* to help spread the word. Thank you for your support.

A New Human Story: A Co-Creator's Guide to Living Our True Potential. Copyright © 2021 by Sonia Molodecky. All rights reserved under International and Pan-American Copyright Conventions. This book is sold subject to the condition that it shall not, by way of trade or otherwise, be lent, re-sold, hired out, or otherwise circulated without the publisher's prior consent in any form of binding or cover other than that in which it is published and without a similar condition including this condition being imposed on the subsequent purchaser.

All proceeds from this book will go towards youth empowerment programs.

CONTENTS

Welcome! 10
Foreword 15
Introduction 19

PART 1: A New Vision for the World   25

The Story of Getting Back to Me 29
Yes, It's a Crazy World Out There! 48
Our Opportunity 55
We Are Energy Beings (*a.k.a.* Superheroes) 73
We Are All Infinite Co-Creators 90
We Are Nature 103
Creating Heart-Centered Realities 121

PART 2: The Co-Creator's Toolbox   161

From F U to Bless U 162
Connect to Create 165
Set Your Default to the Positive 170
Tune Up! (*a.k.a.* Energy Hygiene) 176

Zip Up to Keep Up	184
Grow your Roots	186
Smile!	187
Forgive Yourself	188
Practice Reciprocity: Let the Energy Flow	190
The Age of Celebrity Is Over. Sing Your Beautiful Song!	195
Commit. Discipline Makes Superheroes Out of Everyone	198
Get Up and Move!	202
Humor	204
Believe in Magic, Create Magic	206
Get Your Hands Dirty in Creativity and Imagination— the Portal to Your Higher Self	208
Heart-Centered Warriorism—Get out There and Create!	212
Gratitude & Acknowledgments	218
Bibliography & Resources	220
About the Author	223
Biographies	224

This book is for people of all races, religions, ages, sexual orientations, nationalities and creeds. We are all one Earth family.

It is especially for the youth, who are the true superheroes of our New Human Story. May the messages in this book inspire you to stand in your truth, your power, your beauty, and let your own unique brilliance shine bright. It is your inner essence that will give you the wings to fly.

And know that we've got your back!

We are born of love and will all go back to love. We spend our whole lives looking for love—to be loved, to feel loved, to give love. Yet we fear going deep into our hearts and allowing ourselves to fully feel and express the pure unconditional love that lives there. We fear standing in our love—our greatest strength and power. It is the greatest force in the Universe, and it is our greatest weapon as Heart-Centered Warriors. Darkness, fear and hate have no chance against love.

— SONIA MOLODECKY

NOTE

Throughout this book, I use reference to *the Universe, Source Energy, Great Spirit, The Everywhere Spirit, The First Creator, God*. This is not in reference to any religion or deity. These are all used interchangeably as the first-source energy—pure love—that flows through all things in the Universe... including us humans!

I also reference both the terms First Peoples and Indigenous as having the same meaning, because different cultures and Nations have different experiences with these terms.

WELCOME!

You are wonderful just as you are. We are told this but we don't believe it. We think we have to seek out gurus and answers outside of ourselves. Our core being in each one of us is divine beauty and perfection. In fact, we have limitless potential—superheroes have nothing on us! The things that we are capable of are beyond the realm of what we can currently imagine, and we have stopped believing.[1] We have allowed many layers to be put on top of us that have dimmed our bright light. In this state of camouflage we are operating under, we have been duped into believing that we are *less than* and we agreed.

---

1   Yes, just like Peter Pan, we must believe again, and then we will quite literally be able to fly, to manifest food, to clean the oceans, to create a world that is as it's meant to—full of vibrancy, healthy life, creativity, joy and abundance. But please do not try to fly right away. We have to retrain our minds to believe in our own potential before we can advance to these levels. Start with something a bit more... grounded please.

But the truth is, we have been in the dark ages for a very long time. We have hidden away our light and built a society around us that reflects that fear, that darkness, that suppression of creativity and the divine in each one of us. A massive shift is here and an opening up of those energies. This is the greatest moment in the history of humanity to be on this planet because it is the reawakening of the human spirit. All that means is that we are ready to start to take those layers off, in order to create from a place of peace, love, joy and regeneration, instead of fear, hate, anger and destruction. That is all that is needed for us to collectively build a world that is healthy and thriving for all. We don't need to change who we are or be someone we are not. All we have to do is take off the layers and become our true selves again. We will all rise and shine. And we all have front row seats to the greatest show on Earth. So, get ready!

The dark will put up a fight, like any emotional toddler that can't get its way. There may be fear and hate being revealed, control mechanisms that attempt to suppress who we really are and policies that impact our basic rights. How do we survive these crazy times? Trust your intuition, that quiet voice deep inside. Find tools to stay centered and grounded. Keep your mind strong and healthy. Begin to make your own decisions and develop your own discernment. Think for yourself and question what doesn't make sense to you. You are right. Do not be afraid. Your intuition is always right. Trust it. Let it lead you. Let it guide your decisions. You are far greater than what we are told.

Love is the strongest force in the Universe, and we are all beginning to remember the beauty that is within us all. This force that can build Universes and tiny microbes, is inside each one of us and all around us. Quan Yin, the goddess of divine feminine energy reminds us that this is a time of 'thinking' with our hearts more than our heads. It starts with loving ourselves and opening our hearts to be our guides.[2] We are now all becoming explorers of ourselves. It's time to get to know who we truly are and accept it all. We will have crooked toes and twisted noses, strange sense of humor or awkward social interaction. It is all part of who we are and it's time to embrace it all!

We all have the opportunity to graduate from grade 1 to grade 2 (because our Earth Mother is upgrading to a new, higher frequency and taking us all with her!). But we will evolve and grow as quickly or slowly as we individually choose. Some are choosing to stay in grade 1 because they were so good at it.[3] But those skills and abilities will not serve us in grade 2. There are higher states of being and therefore responsibilities to act accordingly (*a.k.a.* self-governance). The school we are in is actually quite grand. It has many, many rooms and we have been fighting over *one*. Let's get out there and explore the others. There is so much more to experience. There is also a great deal more wonder, excitement, beauty, magic and awe. We have only experienced a tiny drop in a vast ocean of our own limitless potential, never mind what we could achieve if we actually worked together!

The systems we have created have never worked for the well-being of all. No matter where in the world we are, we are each experiencing a different aspect of a human experience that is in its early stages of conscious evolution. Meaning, kind of a shit show! And with all the trauma and karma being pulled out of the dark corners around the globe, we are seeing it with fresh eyes as if for the first time. It will become obvious and the signs will start to show as the fog lifts. Trust it when it does. Allow the darkness to be released with love and the light that is inside of each one of us to shine through. There is no darkness that is possible when we stand in the light. We are all being rewired

---

2   Shih Yin, Spiritual Minister of metaphysics and energy healing, and channel for Quan Yin, the Goddess of divine feminine energy.

3   Like Billy Madison.

and upgraded for this new reality and it can be uncomfortable, painful and just plain weird. But the only reason that the outside world has the power to say we can't be our true selves is when we believe it to be so. We don't need to go to waterfalls in the Amazon to find our powers.[4] They have been inside us all along. We are moving from the age of gurus and teachers to being our own gurus and teachers. Trust your intuition and keep moving forward in the direction of your inner guidance. We are moving to higher states, not lower. From polarity consciousness to unity consciousness. It will be the greatest ride of our lives. Trust yourself to take you there.

This book is for those who know deep down there is more to this human experience than we currently are able to grasp. For those of you who are searching for your purpose and way to make an impact on the world with the unique gifts you have to share, even if you don't know what those gifts are yet. For those who want to have hope again in a better future, but not sure how to make it so. For those who are tired of feeling stuck, feeling exhausted, feeling disconnected from life or feeling like you don't matter in this reality. For those who want to be Heart-Centered Warriors and know deep down that we have limitless potential to co-create[5] the world that we want to see. And for those who are looking for some simple tools to survive and thrive in a crazy world!

I wrote this book because I had enough of the fear-based messaging, seeing people stop believing in themselves, honoring the diversity within the unity of our collective

---

[4] Although they are incredibly beautiful and magical! This is what I believed years ago—that I had to go there to get my powers. That something outside of myself was going to give them to me. It was an incredible experience, traversing jungle and steep rock faces to get into the pit of the waterfall and look through the portal to ask for my powers. But what that journey taught me was that my superpowers were inside me all along.

[5] To "co-create" means to manifest with Source directly!

family, and I wanted to see what could happen if we all got together and began to imagine something better, to intend and co-create something better, from our limitless potential, from our heart-centers—our true essence. What would that world look like...?

FOREWORD

By Arzu Mountain Spirit, Elder and traditional healer of the Garifuna Indigenous Nation, and Founder of the Wagiya Foundation.

In 1991, I was called in "dreamtime" to return home from the United States to Honduras where my grandmother lived. It had been 23 years since my four sisters, one brother and I together migrated to the United States from Central America, to be reunited with our parents.

I took some vacation time and made arrangements to go. The trip turned out to be one of the happiest experiences in my life. The land greeted me and I felt centered and strong. And at my homecoming ceremony in the family "Dabuyaba", the ancestors greeted me and made a request of me that, at the time, seemed insurmountable and unreasonable. But I complied and I agreed. And that request was to take the lead in helping our people return home; help our people to heal. To our ancestors, healing is connecting to the land.

They said they would provide everything I needed; I would just have to surrender to the request. I would not have to worry about food, shelter or clothing. They would provide everything. They would send what they call "friends", to help me and to help us. They kept reassuring me that I would not be alone. That they were sending me helpers, and these helpers would be descendants from all Nations of the world. And that they would have foreign accents and be mostly European.

They would be working to balance the scales of justice to restore harmony amongst Nations. They would be very young. All of them would be under 30. I would feel them. They would seek our medicine and they will heal from our medicine. They said they were here already because they were all already spiritually connected. They told me that I will see proof of their work in the world because they would already be doing what we need them to do. They will be the kind of people who will do as they say and say as they do.

But how would I find them? All I had to do was return home, they said. And once I returned home, all my needs would be provided. I would have a home; I would have all the food I needed and all the clothing I needed. I would be taken care of. So, my instructions were to return home and set up a healing practice offering only the medicine of our ancestors, which I was happy to do. I did not have to look for them. All I had to do was go home to Belize.

And it was in Belize where I met Sonia Molodecky. Shortly after having set up my practice, at my door appears this skinny little white girl filling all the criteria that the ancestors had given me. She shared her vision of what she wanted to do with her life and the organization that she was creating. She was articulating all the criteria that the ancestors had given me. The more she shared with me, the more I understood that she was definitely a helper but very different from the rest.

Sonia was unique. Out of all the helpers I met prior, she was the only one whose mission it was to take on the task of healing Indigenous Nations across the globe. She was not offering a band-aid cure like other helpers in the past. Sonia was offering holistic Nation-Building and economic solutions that were relevant to the Indigenous way of life and with respect for our Indigenous modalities. She was already spiritually connected, and at this tender young age she was already walking her talk. It was unbelievable.

But what was I supposed to do with this child that they sent me to fight an adult war? I didn't believe that she understood the magnitude of what she was suggesting be done. I wasn't sure she understood, but at the same time, I did not want to discourage her. Very quickly, I was happy to be proven wrong. What appeared to me as a disarmingly humble little girl, turned out to be one of the most brilliant minds and beautiful human beings I have ever known in this life. And to be so young.

I didn't know how to define Sonia. She told me that she was an attorney, so I embraced her as an ally. And it quickly became evident that she was not that at all. To Indigenous People, allies have a history of betrayal and do not remain in your corner without guarantee of personal gain or benefit. Allies take sides out of convenience, so they can be dishonest and will betray you for a greater gain. And that was definitely not Sonia. I then defined her as a partner. But the term, again, was not broad enough to describe what Sonia was unfolding before me. Partners stay in relationships because of personal gain and what they gain from each other.

Through phone conversations with her in Canada and I in Belize, Sonia kept giving us time and attention and very good advice. And through the years, she never asked for anything in return. She just wanted to make sure that we were okay and that the right decisions were being made, always asking if she could help. She cared for my people the way I care for my people. And I grew to care for her in a way that required redefining the relationship again. The only word left was "friend"—that person who is there through thick and thin. Someone you can turn your back on and will be there when you turn back around. Someone who has your back no matter what. They watch out for you and they ensure that you're not in danger. This is Sonia. The person who will never purposely lead you into making decisions that are not good for you. This is Sonia. A true friend. Someone who I know will always have my best interests at heart and the best interests of our people. She's one of those people who are honest with you and they do what they say, and they say what they do. And they show up just because they said they would. That is Sonia Molodecky. My true friend.

It was important for me to return home so that I could meet Sonia and the helpers of humanity that she keeps bringing into the global Indigenous fold. God bless her. These are kindred spirits busy building bridges to ancestral wisdom, transcending all boundaries between the Nations of Earth. The ones who are walking the talk, with her. She finds them and they find her.

Our organizations are extensions of our individual life missions, to help Indigenous Nations across the globe return home; to heal our relationships with the Earth; and to balance the scales of justice. The Global Indigenous Development Trust and the Wagiya Foundation are true friends, because of the relationship that has grown out of Sonia and I meeting. Together, we are greater than the sum of our parts, and we are helping to heal the wounded space between all Nations across the globe. Thank you, Sonia.

Just to reiterate, I, Arzu Mountain Spirit, on behalf of myself and the ancestors, am proud to call Sonia Molodecky my true friend.

INTRODUCTION

I was 28 years old and working as a young ambitious lawyer at the top corporate law firm in Canada. I was involved in deals that people read about on the front page of the newspaper. I had a beautiful condo in an exclusive designer boutique building, in the most desired part of town. I had an expensive wardrobe that filled my walk-in closet, and I ate at all the best restaurants in the city. I travelled the world extensively, from Argentina to Egypt to Thailand to Spain, and places in between. I had it all. Or so it appeared.

Deep down I was miserable. I felt as though I was living someone else's life. I'd put on my suit and ride my bike the 6 blocks to Bay Street in Toronto's financial core, take the elevator to the top floor of the tallest building and walk into my spacious office, close the door and begin the 18-hour day of running multi-billion dollar deals. Stress and panic attacks (which I experienced frequently) were, of course, not allowed, so I'd close the door in the middle of the day and quietly cry, vomit, and then clean myself up, put a smile on, and go back out there. 'Work hard, party hard' was the mantra, and I lived up to it well.

But the stress was literally killing me. I was getting low-grade fevers for weeks on end, running on steam, and popping Advils like it was candy. All the while, I could not ignore the grave injustice, pain and suffering all around me. Throughout my travels as co-chair of my firm's Latin American practice, and prior to that my work in human rights and political advocacy for the UN and local governments around the world, I saw first-hand the conflict, marginalization and pervasive discrimination that existed, particularly with respect to the First Peoples. I also saw the utter environmental destruction in our endless pursuit of growth. I knew in my heart there was a better way, and I felt like a hypocrite, 'enjoying' life while others and our natural world suffered so greatly.

But what was the solution? And shouldn't I be happy to have a great career, loving family and friends, a supportive partner, and financial stability? I was frustrated with the state of the world and felt helpless to create meaningful change. I didn't know the answer, but I was relentlessly searching for a way that was rooted in the values of who we want to be as a people, our relationship with each other, the Earth and all beings. A way that had a shot of continuity on this beautiful planet. Yet I could not seem to find what truly *worked* for people and the planet. And I really didn't know where I fit into all of this. What was my purpose? Was I going to be given a mission?

Then in 2013, I met a former Indigenous Chief and now Elder, at a conference in Vancouver, Canada. He shared how he and the other leaders at the time led in the transformation of his Nation, building a strong and vibrant economy, based on impeccable values and principles of who they were as a people and their responsibility as stewards of their lands. He spoke about the economy being a means to a healthy and prosperous Nation, not an end of itself; about reinvesting back into training their people and building natural leaders; about ensuring they were decision-makers in how their lands and resources developed; about how they evaluated projects based on environmental, social and economic merits, in that order, but with the economy ultimately being of service to their people and their role as stewards of their land and water; and about the environmental award they won for re-routing a road to protect migratory paths of wildlife. They were not against development but authors of what it looked like. And it worked.

Thirty years later, they had brought pride and dignity back to their Nation and were thriving, as was their environment and their next generation.

I was blown away! A model that finally made sense to me, and it was right here in Canada, with an Indigenous Nation. It was possible—sigh of relief! It may not have been perfect, coming out of a battle with abject poverty, oppression and marginalization, more than 35 years ago. But it was the best I'd seen and it just made sense. It was a place we could start from and tools that others could use to build their own versions based on their own vision and values.

We spoke over the next couple of days about sharing their story with other communities around the world, to show what was possible, supporting the development of natural economies, and sharing tools to empower people to empower themselves. Two weeks later, at a breakfast diner in Vancouver, we sketched out the business plan for the new organization's mission and vision on the back of a napkin: The Global Indigenous Development Trust was born. Later that day, I walked into my law firm partner's office and quit.

We have since visited and led workshops with communities across the Americas and have been invited to countries across Africa and Asia, and to Australia. We have been to heights of 5,000 meters in the Andes, into the heart of the Amazon and the Central American rainforests. And now we have come full-circle back to Canada to support Nation revitalization at home.

Being of Ukrainian heritage (as my first language) and having grown up with very patriotic grandparents who were part of the anti-colonial movements during the Soviet era, I learned from an early age the devastating impact of genocide, colonialism, and losing one's lands, identity and way of life; the difficulty of overcoming imperial-colonial legacies; and the importance and value that came from being rooted in culture, language and tradition. What at first seemed like a disparate career path became both a professional and a personal journey of understanding and coming back home to myself.

Since then, I have been to the heart of darkness in South America's mining industry; in the middle of conflict zones working to bridge peace; been the source of targeted hate for standing for what is right; witnessed extreme poverty and destruction; traversed jungles and swam in the rivers of the Amazon; lived in remote villages; participated in revolutions and democratic elections in Cambodia, Egypt, Mexico and Ukraine; defended human rights; set up businesses in a multitude of foreign jurisdictions and learned to speak three languages fluently. I became a kundalini yoga teacher, a shaman apprentice, an energy healing facilitator; I worked with medicine men and women from around the world; I practiced many forms of meditation and felt the vastness of the Universe within myself; I sat in ceremonies across the Americas, communicated with plants and trees, practiced sound healing, hiked one of the Seven Summits, and swam open water marathons; and, above all, I have experienced the sheer beauty, wisdom and love that exists within our connection to Great Spirit/Source Energy/the Universe.

These experiences have given me glimpses into who I really am, who we all are, and how powerful and beautiful we truly are. I definitely didn't believe it before. But the more 'magic' I experienced, the more I started to wonder, "maybe it is the norm, not the exception?" And I started to truly believe in our non-separation from all that is and the limitless potential that exists within each one of us (no exceptions).

And so began the journey of getting back to myself. In that process, I became a true explorer of myself. In the years that followed, I sold my fancy condo; shipped my expensive clothing off to places in need in Ukraine, and donated what I no longer needed to shelters in the city; and I ultimately left my marriage. I was left naked—without titles, designer names, or a prominent career—starting from a blank canvas, but one in which I was the master painter.

Through my book, I share stories about our true human potential, as well as some tools for believing again in what's possible, for getting back to the heart-center (*a.k.a.* our true self), and for reigniting our imaginations to co-create an incredible and thriving world *together*—from the inside out, from our true essence. I hope my experiences (and some lessons of what not to do!) inspire people to have the courage to come back to themselves (if I can do it, so can anyone!) and become active participants and co-creators of our collective story. I believe this is the challenge of the next decades and also our greatest opportunity.

What will our New Human Story be?

A NEW VISION FOR THE WORLD

I see a world driven by the power of love, not fear. Where human beings treat each other with *humanity*. Where compassion, kindness and generosity of spirit are characteristics we teach in schools and strive to embody in all we do. A world that values nutritious food, clean water and life-giving air, as possessing greater inherent worth than gold and silver. A world that creates abundance rather than scarcity by encouraging each individual to shine their true light and share their inner gift with the world, creating a creative economy infinitely bigger.

A world where the true worth of an individual is not measured by numbers but by who they are as a person and the beautiful energy they contribute to the world. A world that rewards the good rather than punishing the bad. A world where currency is a reflection of true value created, not a stealing of our time. A world where the water heals, the Earth gives life, the air purifies, and we align with it and allow it to do so. A world where all humans have enough food, water and daily sustenance to thrive. Where energy is free and comes from our own internal force of natural power by tapping into our higher frequencies and using them to power a healthy world.

A world where we live in reverence and have a reciprocal relationship with all beings. Where we are grateful for the daily gifts we are given by our nature relatives and the nature that we are a part of. And in this gratitude, we create greater abundance. Where we remember our true power as divine light beings—energy masters of the highest order—co-creators of our Universe and beyond, each a micro Universe connected to the greater one whole that is all of existence. Where we remember our own inherent ability to heal all illness and pain. Where we remember the state of joy, that we have the right and the ability to choose.

A world where magic is the norm, not the exception. A world where we believe in cooperation and healthy competition, within the limits of natural law. A world where everyone is busy with life-affirming and personally enriching work. Where each of our own genius is contributing to making the vibrant whole, as is nature's brilliant intelligence, our brilliant intelligence. And where each of us is valued for that contribution to the whole. A world where a smile is worth more than a frown. Where 'smart cities' become 'healthy cities' and '5G' becomes '5D'.[6]

Where our minds are strong and we manifest powerfully. Where our houses are healthy and nurture our soul because—like nature—they breathe, self-regulate, and there is no waste, instilling a sense of calm and ease in our bodies. Where we look each other in the eyes and

---

6   5D—Fifth Dimension. It is said we are currently living in the Third Dimension which is a reality we experience only through what we can see, touch, taste and smell. We are opening up to experience a higher realm of consciousness that includes multi-sensory perception, experienced through the heart-center—through sensory perception (*a.k.a.* those icky things we refer to as "feelings"). This is also where all the magic happens!

communicate our inner power through our natural sensory preceptors network.

Where AI (Artificial Intelligence) supports HI (Human Intelligence). Where we remember that we chose to come to this living library to learn and experience life in this human form, and that we are energy beings living a human existence. That each day is a gift of infinite learning and possibilities. A world where we collectively choose to take back our collective story. A world where we no longer allow limited thinking and limited being to write our collective story for us.

A world where we are active co-creators in the world that we want to see and be a part of. A world where everyone is happy, healthy and prosperous. Where we live from our heart-centers and are guided by our higher-selves, our true selves, in all that we do. We are no longer "drunken monkeys".[7] We are true spiritual warriors of the highest order. Ours can be a new human story grounded in truth, in love, in higher understanding. This is the vision I see for the world.

What is your vision?

---

[7] Ram Das referred to our minds as "drunken moneys"—not able to sit still, focus or concentrate. As a result, our minds become weak and it becomes difficult to find peace or consciously manifest our reality.

• INVITATION •

We are energy beings.[8] We are interconnected. We are one.

We know how to create stars and planets, trees and oceans. Because we *are* the stars, the planets, the trees and the oceans. We *forgot*.

In these times of great uncertainty, challenges and shifts, it is critical that we wake up to who we are and our ability to co-create the world that we want to see and be a part of. Let us not hold onto the fear and the hate. Acknowledge it and then move towards the beauty that is within each one of us. Let us stand in the truth. Let us not be afraid. Let us all sing our own beautiful song.

This book is not intended to be read sitting down. Get up, get your hands dirty in the Earth, listen to the birds chirping and sing along, observe the nature around you that you are a part of, then breathe life in and bring forth your own true nature. Dance and, most of all, smile because you are an incredible force of powerful creative energy that lights up the world! And that is what is most needed at this time. It's you—*your* light! So, bring it on!

---

8 We are all energy (pure love and light) living a physical human experience. That is our core essence and makes us all energy masters in the making.

THE STORY OF GETTING BACK TO ME

When I was very young, I loved playing in the dirt, painting with my hands, diving into the water, and exploring the world around me. It felt natural to be able to communicate with animals and plants. I knew we had a connection with all life beyond what we could see, touch, taste or smell. As a child, I drew and wrote about the natural world and our place in a more harmonious existence. My heart was open to all of creation. I saw the light in everyone and was eager to shine my light out to the world. Life was pretty good for a four-year-old!

And then reality hit, as in all of our journeys, with economic hardships that make life immediately challenging. In our case, it was when the housing bubble burst in Canada which sent those working and investing in it into a tailspin overnight (and that included us!). My parents worked hard and did what they could to protect us from the stress of surviving the storm and putting the pieces back together. We had good years and tough years. I remember going to fancy designer kids clothing stores with my mom and sitting in the change room as she would sketch out the dresses we'd try on, and then go home and sew them for my sister and me so we would have pretty dresses to wear to church.

We also got to spend a lot of time with our grandparents, which we loved. My grandparents came to Canada from Ukraine as "displaced peoples" after the Second World War. They fought for their sovereignty, their lands, their language, their culture, but ultimately had to leave their land and all they knew. As a result, we grew up with great patriotism; we were taught to fight for what we believed and to always stand in the truth.

Ever since I can remember, I had a desire to make this world a better place. I was so passionate about social justice from an early age. Yet, as a young child going out into the world, it was so much harder to navigate. I felt the hatred, the greed, the anger and the fear. In those times, "empathic or sensitive children", "energetic boundaries", "energy vampires" were not really known. So, I struggled through it, and my way of dealing with my sensitivities was to disconnect. I spent much of my teenage years literally out of my body, not sure who I was and not at all understanding how this "vehicle" (my body) worked or what to do with it exactly.

The world was a harsh place, and I was constantly feeling the pain and the devastation of the world. It was brutal out there! As a result, I took on a lot of anger, and became an activist early on—about the environment, about human rights, about animal rights, about anything that seemed to me to be an injustice in the world, and there were a lot to choose from!

I WENT TO LAW SCHOOL TO CHANGE THE WORLD[9]

As a young idealist and human rights advocate, I spent time working for the UN on democratic reform in Ukraine,

---

9   As any smart and idealistic kid did—basically 90% of my first-year class.

I worked for the Public Defenders' Office in Buenos Aires, and I worked for the President of the Human Rights and Justice Commission in Mexico. What I got was a big awakening about how the world really worked and how little power it seemed I had to actually affect the change that I so desperately sought. At the time, it appeared that the corporate world had all the power, and I thought there was a way there to affect change.

So, I went to work for the largest corporate law firm in Canada. For a creative kid, it was a big shock. My first day, I walked into the plush offices—with purple velvet couches, white marble floors and famous artwork draped on the walls—and felt like I was in someone else's dream. But it was a new experience, one for which I was committed to giving a go. I worked very hard: I implemented Corporate Social Responsibility practices, opened up new markets, and I learned the corporate world well. But, try as I might, I could not seem to implement the changes I was really seeking in my heart.

IN THAT TIME, I LET A LOT OF FEAR INTO MY LIFE

I began to fear life itself as I became more and more disconnected from myself and the beliefs and reality I lived in as a child. This fear shut down my heart and allowed me to experience my thinking mind[10] in all its power and strength. I remember going to a reiki master for a tune-up and balancing of my chakras,[11] and afterwards she said to me, "You have put bolts all around your heart so the energy is not flowing there". I guess in my young mind, I was protecting myself from the "big, bad, cruel world", but it was also preventing me from

---

10   As contrasted with my higher mind, which is all-knowing and connected to Universal Source Energy. The thinking mind is good for calculating, logic and rationalizing, but is limited.

11   At the time, this was my weekly "hit" that I came to depend on to get me feeling good... enough to keep going at the pace I went at that time.

living in balance, living from my heart and from connecting to life, to people and to myself in real ways.

Pretty soon it began to eat away at me—being cut off from my heart-center, from my life-force energy—and it nearly destroyed me. You see, in my rage and despair at the state of affairs in the world, I had forgotten about my true superpowers.

I started putting an intense amount of pressure on myself (*a.k.a.* beating myself up—why we do this, I don't know!). As a Virgo, to say I am a perfectionist is the understatement of the century. When things didn't go as I had so carefully planned, I would beat myself up and get really angry with myself. I used to get these awful low-grade fevers all the time. They would literally go on for weeks. They were wearing me into the ground; *I* was wearing myself into the ground. I didn't understand in my early twenties that I was *pushing myself* instead of *flowing* with the natural rhythms, thus completely cutting myself off from the Universe's life-force energy. I was running on exhaust fumes. This went on for many years, well into my thirties (I'm embarrassed to say!). I was all drive and no flow. I'd put a smile on my face, so everyone thought it was all good, but I was driving an agenda and doing it without my power cord.

I started getting lots of infections in my lower abdomen. My doctor sent me for an ultrasound which revealed incredible scarring on my uterus from all the fevers. I was told that if I didn't get it under control, I wouldn't be able to have kids. I was 27 years old. I realize now this stemmed from insecurities of not feeling good enough and needing to be the smartest, to do the best, to excel at everything, to prove my worthiness. I let the outside world dictate who I was and how worthy I was.

SOMETHING NEEDED TO CHANGE

So, I did what any 30-year-old would do: I went to Google and typed in "stress management". I looked and looked and nothing resonated. Then I found it: "Stress and Vitality—Kundalini Yoga". This class resonated—a kundalini yoga master and a homeopath teaching together, the spiritual and the natural medicine. This was for me! I signed up and put it in my calendar, too busy to read through the program.

On the day of, I showed up ready with my notebook to learn. I sat down in the front row (yes, I'm kind of a nerd), and the teachers introduced themselves and said, "Welcome to Stress and Vitality Teachers Training". Wait... *what??!* Teachers training?? What did I do! I didn't see that anywhere! Is there time to run? I am not ready for this—I can barely keep myself in check! But the Universe works in wondrous (and often hilarious) ways and it turned out to be exactly what I needed. For the next five days (eight hours a day), I sat there, did the practices, exercises and meditations, and got my kundalini teacher's certificate in Stress Management and Vitality. I could feel the Universe laughing. Now, I had no excuses.

One of the homework assignments after the course was completed was a 90-day mantra for 31 minutes each morning.[12] I could not miss one day or I'd have to start over. I was committed. It was right around this time that I was working in South America with Indigenous communities and the mining sector. I got pulled right into the nightmare of horrors that exist in some of these places. I was a bright light that was new, naïve, and wanting to "save the world", which made me a target. There were nights that I only slept a couple hours, but I would get up and do my mantra. It was the discipline of doing this mantra and the grounding I received from it that, I believe, saved my life

---

12 Mul Mantra, one of the most powerful mantras and said to be the mantra of enlightenment. It is said that this powerful mantra contains the core, essential truth of creation, and its vibrations are so powerful that it can change your fate and even help you "rewrite" your destiny. For me, it helped to get rid of negativity and find greater happiness and peace inside. Look it up and give it a try!

during that time. It kept me centered (as much as I could be at the time) and grounded in myself, so that I could make smart and safe decisions.

After that, I started going to a restorative yoga teacher regularly to bring my adrenals back to life, which it seemed I had completely destroyed from all the stress. She looked at me and said I was all yang and no yin.[13] Where do you find this thing called "yin" I wondered? She told me point blank, if I'm going to have an intense career, I'd have to have more balance in my other activities (at the time I was doing Muay Thai martial arts training, swimming training, and doing ashtanga yoga (the most physically intense form of yoga and quite hard on the body). Oh, and of course, in my corporate law world it was all "work hard, party hard", so regular nights out drinking hard were the norm. Okay, so I *was* a bit intense, but I liked it. Only problem was, it really didn't work for me anymore. My body was falling apart and screaming out for help.

MY SISTER WAS MY INSPIRATION FOR WHAT I DID NEXT

My sister has always been extremely strong and disciplined in all that she does, which I have always greatly admired. It was through some of her own experiences that I was inspired to look at how I was treating my body and to connect with what it needed to feel healthy again. Part of that involved cutting out all alcohol, caffeine, gluten, dairy, and eating organic natural foods—permanently.

At first it was brutal! My body was purging years of abuse and I went through mood swings, night sweats and all sorts of brain fogs. I remember my sister and me going

---

13   An ancient Chinese philosophy that talks about balance existing between the yin and yang—*Yin* being the feminine, the allowing; *Yang* being the masculine or the pushing energies.

to restaurants and we'd order just plain veggies with no sauce or butter, and then I'd order a glass of red wine, just to smell it! (LOL, it is just like planning a vacation where they say that half the enjoyment is in planning the trip. In this case, half the enjoyment was in the smelling of the wine!) Eventually, the toxins began to clear out and wow! It was transformational!

My energy started coming back, and I could feel my body re-forming to a new normal. I started getting the motivation to do daily meditation, yoga and outdoor walking. And that eventually created enough stillness and balance within me to start making decisions clearly. It eventually led to a big shift where I jumped off the cliff: I quit my law job and started the Global Indigenous Development Trust with an incredible Indigenous Elder. Now that I could feel myself, I could no longer deny myself and my heart.

While I was enjoying my new career and feeling passionate about what I was doing, my default setting was still set to "push and achieve". It was easy to remember my "new way" in good times, but when things got busy and stressful, I'd default to what I knew, what I was comfortable with. And it was not so easy building our organization! We were starting an Indigenous organization at a time when "reconciliation" was an unknown term, when people didn't know what they know today and definitely were not open to the vast knowledge that the First Peoples could contribute to building a better world.[14]

---

14 Other than the pharmaceutical industry, of course, which has known this for decades. It is said that the majority of our modern-day medicine derives from Indigenous knowledge around the world, most of it never being acknowledged nor compensated. Yet, the First Peoples continue to share, for all of us to heal and for the benefit of humanity.

> *I would like to right now pause to acknowledge and say thank you to the knowledge systems from which we have all benefited, from which our parents and grandparents have benefited, over the years and at great expense to these incredible Nations. It should not have happened as it did. It was wrong. I apologize for the greed of humanity. Thank you for the love you have shown. We can all learn a lot from these selfless acts of generosity and unconditional love.*

At the time, I had people hanging up on me, canceling meetings, and it was a real challenge to get basic funding for our work or community-led projects. No matter what I tried, I simply could not get anyone to listen or care. It was such a frustrating experience—knowing there was a better way, tremendous contributions to be made, and being made to feel invisible. As a white Canadian female, this gave me a very tiny glimpse into how the discrimination the First Peoples face, might have felt. Yet, I will never truly know what that must *really* feel like, for I was simply a visitor to that experience. To this day, it saddens me beyond measure to think that this is an experience that people in our human family have to face *each day*. For, all any of us really wants is to be seen. Witnessing this discrimination only served to push me to work harder.

SOON ENOUGH, THE FEVERS AND SICKNESS WERE BACK

I tried everything and not just tried, but I did it the Virgo way: commited with discipline, did it with conviction. I started practicing kundalini yoga and restorative yoga;

going to my naturopath and homeopath regularly; working with shamans, monks, medicine men and women; reiki; practicing meditation; trying plant medicines and vitamin IV infusions; Chi-Guhn, Tai Chi, traditional Indian medicine, hypnosis, sound healing, Neuro-Linguistic Programming (NLP)—you name it!

I remember going to an Ayurvedic doctor, and she tested me to see what combination I was of Pitta, Vata and Kapha. We are each a combination of these Earth elements, she said, and it's about finding the balance (wait: I am seeing a theme here!). I was 98% Pitta, meaning I was quite literally on fire! I used to be so caught up in work and how busy I was—from being a corporate lawyer at a major Canadian law firm, to starting a global organization and being an entrepreneur—that I felt I didn't have time to stop and meditate or even simply sit still. But I had to just do it or I would crash and burn yet again.

So, I'd stop what I was doing in the middle of my workday, take my yoga mat to the park in the center of Toronto's downtown core, and just sit and be still. Well, at first it was painful. Five minutes was a tough sell. But I committed and did it. What I discovered was that my afternoons just took care of themselves. Work somehow got done: people would call that were on my 'To Do' list and tell me the task was already done! This was the Universe telling me that I was on the right track and to keep it up. The less I "did" and the more I "just sat in stillness, being", the more efficient my days became and the calmer and healthier I became.

Around this time, I started going to see a spiritual advisor. I didn't really know why but I knew there was more I needed to learn and I was still so disconnected from myself that I didn't even know what I needed to do other than to get some guidance and help. I still felt as though a part of me was acting out a part in a play and that I was not actually leading from my soul (not that I even knew what that was at the time!).

I remember her asking me at our first meeting, "What are your favorite things to do? What truly makes you *happy*? Like when you are doing it, ahh, you are at peace, you are in bliss, and you feel HAPPY?" I looked at her, paused for a moment, then started to ball my eyes out. She asked me why I was crying and I said, "Because I have no idea what I like or what makes me really happy!"[15] It was such an easy question and yet, for me at the time, it was like climbing Mt. Everest. I literally had no idea. She looked at me, smiled, and said it was alright... that we would find it.

MY HOMEWORK WAS TO "DATE MYSELF"

I would pick an activity, tune in, experience it in *presence* (not sure I knew what that was yet—aren't we all present and accounted for, I thought?), and write down in my journal how it made me feel. I started with simple things like taking a bath with candles (yes, I am cheesy), going for a bike ride, a swim, a walk, a hike, reading a book outside under a tree, staring at a leaf or an ant walk across the driveway, and all sorts of pretty regular "mundane" things.

At first, I wasn't sure I enjoyed any of it. I mean, where was the adventure and the excitement? What I soon realized was that I didn't enjoy spending time with myself. Ouch! That was a pretty brutal one! And guess what? If I didn't enjoy spending time with myself, why would anyone else? Oh man, this was going to be a rough journey! So, now I had to start to actually *like* myself.

I made so many countless lists of what I was good at, what I was proud of, what my skills were, accomplishments, attributes, contributions. I wrote and wrote, and words came out, but it all felt fake. I felt like a fraud. Like it wasn't me at all. It was as if I was describing some *other* person who was all those things. I could not have felt further from

---

15 And I am not a crier. Sad movies, funerals—not a tear. I had become tough on the outside and inside. But now... all of a sudden... waterfalls.

that character I was describing. But I kept at it, along with daily affirmations, journaling, meditation, writing, and tons and tons of spiritual courses—anything I could get my hands on that would give me the knowledge, tools and skills necessary to tackle *myself*.

I remember when I first realized how negative my mind was. I was trying to get to the bottom of my constant fatigue and "burn-out". I was told to "watch my mind". *What does that mean?!*, I thought. *How do I do that??* When I started to practice it, I was appalled! My mind was negative and worried and controlling all the time! And it was soooo sneaky! I didn't even know I was doing it. But it was making me sick.

So, I set out to reprogram my mind. For six months, every time I'd catch myself thinking or saying something negative, I'd stop, reframe, and say the opposite phrase to the positive. It was a painfully slow, tedious and very humbling experience. For a long time, I remember thinking, "It's not working! What is wrong with me?" And then it started to shift slowly. I'd start to see glimpses and start to have moments of reprieve. And those moments became more and more frequent, until I could actually "see" my mind and become an observer of my experiences. It took a daily commitment and patience for incremental changes.

Breathing was another big one for me. I'd simply forget to do it! LOL! That sounds crazy, but I was a shallow breather. Basically, I'd inhale in the morning and then hold my breath all day and exhale at night. No wonder I was so exhausted all the time! This shallow breathing, I came to learn, is also the cause of added stress and panic attacks, which I experienced frequently. Belly breathing was such a foreign concept for me. The yoga mat I had in my law firm office was used for power naps at 3 a.m. during big merger and acquisition deals and not for yoga or meditation! But I started to pause and breathe. In fact, I had such a hard time remembering to breathe that I had to set the alarm on my phone. It would go off every hour and I'd stop and just breath. Over time, it became a habit.[16] Now it is my go-to for centering myself quickly and effectively.

BUT THE BIGGEST ONE FOR ME WAS CONTROL

Oh boy! For a 'type A' Virgo like me, letting go and surrendering has been the hardest thing I've had to do. I even tried to control the letting go process! I'd go to see my closest friend Lianna who's an incredible intuitive chiropractor, and I'd be locked up tight (stress does a good job of that!). She would get me on the table, and I'd start doing the movements for her! She finally had to say to me, "Can you please let go and stop controlling every move so I can do my job and help your body release?" I just had to laugh.[17]

Fear had set in and I began controlling everything, so my body became rigid. Rigid mind, rigid body. The more I loosened my mind and let go, the more my body opened up. I became more and more flexible as things started to flow, and my body began working so much better. The dis-*ease*[18] in my body started to go away and I started connecting more with my body and responding in real time to its needs.

I'm now able to tell within a few seconds when stress starts to build up. When I start to worry, I now catch it in time and stop everything, take a few breaths, or go for a walk. I practice this ongoing "energy hygiene" so that I

---

16 LOL! Can you believe breathing has to become a habit?! That's one for the history books!

17 In this journey, one thing you need is a big dose of being able to laugh at yourself because we can all be pretty ridiculous at times!

18 "Dis-ease" in the body is a lack of "ease"—lack of flow, in our bodies. This happens in our minds. However, we have cluttered our minds so much that we no longer know our minds or what they are creating. In fact, they are letting off early warning signals: "Help we have hit an iceberg!" and we just can't hear them.

never have to let it build up and have a crash. This was a game-changer for me. When energy is not allowed to flow, when it stagnates (and this can be in the form of control, holding on, refusing to look at something, or ignoring uncomfortable feelings and emotions), we get dis-*ease* in the body.

People have always said that things appear to come so easily and naturally to me. That I can manifest easily, that things "just work out for me", and that I can always manage to put a smile on my face. The thing is, I work at it every day. Over time, I have developed incredible discipline (because I had to!), commitment and once I have a vision for what is possible, I go after it against all odds.

The amount of people who have sat across the table from me after we started the Global Indigenous Development Trust and said to me, "What you are doing is not possible. You won't achieve it". I cannot count on both hands. I shake things off easily and keep walking into the darkness even if it is unknown and full of monsters, quite literally! And I try very hard to not sweat the small stuff (which was more "try" than "not sweating" for a very long time). I put a smile on even if I don't feel it because even that simple act begins to turn my day around. It requires daily work, constant seeking, being open, learning, growing, and challenging myself. It's actually a daily commitment to be a true explorer of myself and seeking the best version I can be at any time.

What drives me now is my unrelenting belief in the potential of humanity and the desire to make life better for people so we can all live healthy, happy and fulfilling lives, in harmony with life.[19] I once asked my business partner, mentor and friend, Jerry[20], about how he was able to get

---

19 Which [harmony with life] actually makes us happier, healthier and more fulfilled. Go figure!

20 Jerry Asp is the Co-Founder and CEO of Global Indigenous Development Trust, and former Chief and now Elder of the Tahltan Nation.

up every day and work towards a vision that seemed so daunting and unachievable—to take his Nation from abject poverty to prosperity. "Just one day at a time," he'd say. He just got up every day and did a little more, lifted up another person, learned a bit more, built a bit more. He often tells communities we work with now, "It took 20 years to be an overnight success".

THIS IS HOW I LIVE MY LIFE

I have worked hard all my life, from passing my lifeguard exam at age 16 (one of six people who passed in a class of thirty), to learning new languages, to getting into law school and getting top grades and a top job, to working around the world, and to becoming an entrepreneur and starting my own organization and businesses. All that took incredible discipline, hard work, commitment to be always improving myself and learning (even if I didn't always know where I was going exactly). And it didn't always work! In fact, I nearly went broke—a few times—and had to learn a great deal of trust and faith in the Universe and in myself. Throughout all that, I struggled greatly with stress and anxiety, self-esteem, insecurities, massive fear of life, fear of public speaking, fear of failure, fear of success—you name it!

Contrary to what some spiritual leaders say, it is not just "one night of the dark soul" that we feel and all of a sudden, poof! We are "awakened!" It is many hundreds of nights of the dark soul. It is a long journey of awakening, peeling back those layers. It goes on for a lifetime. Even when we are "awake", we are still learning and growing and improving daily. Awakening to me is simply beginning to see ourselves as we truly are and beginning to love ourselves as we might aspire to love others. It's opening our eyes to the life around us and actually choosing to be a part of it. It begins with tiny glimpses, which turn into bigger glimpses, until one day you just start to feel more like yourself.

Awakening is a journey back to our true nature. It is something that happens gradually as we start to become aware that there is more to life, more to this reality and more to ourselves than we were taught. I have had to literally take myself apart and re-train myself to a higher place of understanding and a much better attitude. When something isn't working, I investigate. I am always an explorer and always looking for the gold.

And I have had some incredible people who have helped me along the way. As Jerry said in his acceptance speech for a big award earlier this year, "When you see a turtle on top of a fence post, you know he had help getting there". I have been blessed with inspirational people in my life, and I have sought out guidance and support when I needed it. More often than not, the Universe put people in my path, such as Arzu.

Years ago, I was that young, blond scrawny kid, arriving in Belize to support in any way I could. I arrived with Jerry and we had to take this small 6-seater from Belize City to the southernmost city, where the Maya and Garifuna, whom we were meeting, lived. The plane makes stops at all the towns along the way, every 10 or 15 minutes, sort of like a bus. At one stop, a local man—a farmer from the countryside—gets on and sits next to me. I say "hello" and he smiles, asking me where I am going. I told him the name of the town. Then, just as the plane is about to land at the next stop ten minutes later, he turns to me and says, "You should go see Arzu. Here is her address. Just knock on the gate and she'll see you". He hands me a piece of paper with her name and address scribbled on it, as if he was prepared to meet me. I took the address and thanked him as he got off the plane. It was too strange a meeting to ignore.

So, after we checked into our small hotel, I walked the six blocks and stood in my jeans, t-shirt and backpack in front of this lime green house with a white gate and knocked. Out came this beautiful Black woman with these long braids tied up on top of her head, a vibrantly colorful dress, and gorgeous scarf wrapped around her braids. She looked at me, smiled and said, "I have been expecting you". I looked up and thanked the Universe for what was promising to be a very special journey. But first, over the next two weeks, Arzu proceeded to remove all the "dark spirits" from my body, and that was just painful! It has led to a beautiful friendship with which I was gifted because I remained open to what and who the Universe was putting in my path.

THROUGH IT ALL, WHAT I HAVE LEARNED IS THAT THERE IS NO MAGIC PILL

There is no "one-stop shop". You can't buy "your true self" online. And no teacher or guru can find it for you. It takes small steps, daily. It is incremental and it requires consistency. From all the things I have done and tried, the greatest results for me have come from doing my yoga and meditation practice every day, even for 20 minutes; going for a walk every morning; doing daily grounding work (so important now with all the chaos!); and every day breathing properly again (fully and completely).

It has been the small things, done consistently, that have added up to the greatest results over time. But the cool thing is that, just like exercise, you don't see results right away... and then all of a sudden, *bam!* there it is! It takes time to reach the tipping point with anything, but once you do, it becomes you: *you* become you. And, boy, is life easier when you are *you*!

That's not to say I don't still work at it every day. I'm still learning and growing,[21] and it is still something I work at (or consciously surrender to) daily. I have found that having simple tools that are daily "go-to's" have been super-helpful for me. I have a chant that I memorized and

---

21 Or I'd be dead, LOL!

repeat when needed; I do a few sun salutations in the morning to get the energy flowing in my spine; I journal or go for a walk if I feel the energy is stuck (sometimes I can only muster the strength to get to the end of the hall and back and that is okay: I go for that!); and I always go back to my breath.

These are the things that work for me. What are yours? If you don't have yours yet, begin exploring yourself and find what works for you. They can be very simple!

Now I am uncovering and delving into new aspects of myself, such as really understanding energy. I am beginning to explore the other rooms in the house that we have not been exploring in this dimensional reality. And not through drugs, but through my connection with Great Spirit, Source Energy, the Universe, the First Creator, that is within us and all beings, and the understanding that we are all energy beings and we are all one. It is a journey into complex metaphysics, and I am excited to continue being an explorer, which is what we are here on this Earth plane to experience (once we get this "shit show" under control!).

Yes, I still have fear. I think fear is a natural part of being human. But it is about being able to see fear for what it is and respond accordingly. One of the first books I read on my journey was *Smile at Fear*.[22] It was brilliant because it taught me that we all have fear. We either choose to respond to it by letting it control us, or we choose to be in the driver's seat of our life. Most of the fear we actually experience is just a construct of the mind—fear of looking bad, of failure, of not being accepted. We are now moving from a reality where our egos were in the driver's seat to one where our hearts become captain of our ships and our egos become a support system. In love there is no fear. We may not get to a place of no fear in this lifetime, but we can certainly allow love to rule the day.

---

22 Trungpa, Chogyam. (2010). *Smile at Fear: Awakening the True Heart of Bravery*. Shambhala.

Perhaps the hardest part of my journey was the journey to reconnect with my heart, and bring the two into balance—my heart (feminine energy) and my mind (masculine energy[23])—for me to walk in true harmony. This is what we are all being challenged with now, and it is the toughest part of our collective journey. I had to embrace my nurturing, gentle, loving side, and begin to open my heart again. It was where I truly became a Heart-Centered Warrior, for holding the two in balance is what makes superheroes of us all.

I now aim to hold my heart and my mind in balance, letting my heart lead and create from a place of harmony with both aspects of myself. This is what our world is striving for as a collective.

THE EAGLE AND THE CONDOR

There is an ancient Amazonian prophecy that is more than 5,000 years old. Many Indigenous tribes around the world have their own versions that are quite similar. It tells of a time period of 500 years starting in the late 1400's when

---

23 We all have both energies within ourselves.

the Eagle energy (that of the mind, of the aggressive, of the masculine, of the industrial) would rise and almost wipe out the people of the Condor (that of the nurturing, Earth, feminine, heart energies). We know this as the time of colonization, of the Industrial Age, where the feminine—the people of the land and the nurturing energies—were nearly destroyed in the quest for exponential development and growth.

The prophecy goes on to say that at the end of this period, starting in the 2,000's, we would enter another 500-year period when the Condor energies would begin to rise. This new period would create an opportunity for the Eagle and the Condor to fly in the same sky and create peace and harmony for all of humanity. The prophecy says that this is the opportunity but that it is up to us to realize this potential.

Our new human story is one of balance and harmony. It is of the Eagle and the Condor, flying in the sky together and creating peace, prosperity and abundance for all of humanity.

This is the opportunity. It is up to us now to create that reality, together.

YES, IT'S A CRAZY WORLD OUT THERE!

We *together* created a world that does not work for all of humanity. In fact, it does not work for 99.9% of beings on this planet. Talk about an engineering challenge gone wrong! Yes, we all did it. We collectively got stuck at a party gone wrong. We've all been to those, where we wake up, look around and there are holes in the carpet, spills on the furniture, toilet paper hanging off trees outside, and we wonder, "Wowww, what happened?!" Well, we may not have orchestrated the party[24], but we accepted the invitation to attend, and we stayed longer than we should have.

---

[24] It's like the "cool" kids in school who use bullying and the servitude of others for their own gain, because they do a great job of convincing people that what they have is super-awesome. But at what cost? And is it really all it's cracked up to be, or is there WAY more depth, joy and richness in another way? Hmmm, let's see.

We do not produce enough food to feed ourselves, and what we do have is far from containing the nutrition that our bodies and minds require to thrive. We have destroyed our soils, our aquifers are depleted, our seeds are homogenized, and our animal relatives are treated with cruelty as we ingest their fear into our systems. Our water is polluted, our forests are on fire, our fields are monocrops for animal feed, our rainforests are bulldozed. We spend big budgets on putting up borders and fighting one another rather than feeding our children and healing our families.

We rely on centralized systems that are very costly and highly inefficient, at the peril of local economies and natural life systems. We have given up our own self-reliance and self-resilience—skills that we all once possessed to take care of ourselves—to big-boxed stores and online delivery, trusting the ever more complex (and tenuous) supply chains. We put all our faith in the leadership of our countries without questioning or demanding truth or integrity. We elect governments that rule, not serve.

We expect our taps to turn on each morning and trust that the water that comes out is healthy for us, which we do not question or verify. We have no idea where most of our food comes from, how it was produced or by whom, reliant on delivery systems and fossil-fueled heavy transportation. We accept a healthcare system that perpetuates sickness rather than promotes our own well-being. We accept an education system that teaches conformity to the lowest common denominator rather than inspiring sovereign thinkers. We have given away all our personal responsibility to third parties and stopped thinking for ourselves. In times of uncertainty, this is a dangerous thing.

On top of that, our minds are jammed with useless information; we are over stressed with pressures of life that keep us going a mile a minute. We allow the media to influence our thought patterns and emotions which severely dampens our own innate wisdom[25] and our inherent ability to be our own decision-makers. We turn on each other and allow race, belief or color of the skin to dictate our worth.

We have accepted cities that keep us wired and perpetuate stress. We have accepted nature as separate from us in our daily lives. We have accepted the lack of creativity in our careers and businesses. We have corporations that are chasing only the bottom line to appease shareholders[26] and an economic system rotten to the core which steals our time rather than accounts for our true value. As a result, we are indebted up to our ears and running on empty. We have become passive recipients of systems that serve a few, at the peril of many. We have accepted the ease of conveniences over value. Quantity over quality. Technology over human interaction. Disconnection over connection. We have dug ourselves a deep hole.

Many in the spirit realms refer to this time in human history as the "dark ages".[27] Road rage and erratic behavior on the streets is increasing. Violence and crime are all over the news. Some days it seems that everyone has gone mad! From the government to the media to colleagues at work or at school—never mind worrying about massive forest fires, earthquakes, droughts and floods—how do we manage to keep it all together, let alone prosper and thrive?!

───────────

25 Quan Yin as taught by Shin Yin..

26 No, I am not against all shareholders. I am also a shareholder of many enterprises, but with rights and power come responsibilities. And we all right now have a responsibility to think and make decisions from a conscious place that supports healthy continuity of life.

27 While I do believe darkness is simply the absence of light and therefore this is a time of great healing and transition for humanity, we are experiencing many aspects of what we would call deep denser energies so that they may be brought out to the light and cleared from the human collective consciousness.

The world has reached a tipping point where the systems we have built can no longer sustain life. They must be upgraded, which means we must upgrade (how we think is how we create!). Gerald Jampolsky, MD, in his book, *Love is Letting go of Fear*, states:

> *The world that seems so insane is the result of a belief system that is not working. To perceive the world differently, we must be willing to change our belief system, heal our relationship with our past, and expand our sense of now by releasing the fear in our minds. This changed perception leads to the recognition that we are not separate, but that we have always been joined.*

The reality is that the systems we have built never worked for the whole because they are not natural, so they could never last. Nothing unnatural can continue indefinitely. Nature always seeks to rebalance and harmonize. We have built civilization after civilization on foundations of fear and mistruths, completely against the flows of natural law. So, by definition, they cannot continue. When you build something contrary to natural law, it will crumble, as we have seen since the beginning of time. Nothing built on falsities ever lasts. Only what is built on truth can be sustained because truth is divinity and the foundation of all life.[28]

The ancient Indigenous governance systems held truth to the highest esteem. The Grand Councils of the day were tasked with a search for truth, not debate amongst each other for a victor. The Canadian Charter of Rights and Freedoms is marked with the words "the fundamental values and principles that promote the search

---

[28] Clan Mother Pimastan reminds us of the need for humanity to stand in truth. It is important to not shy away from the truth or be afraid of it, but to look it head on and accept it as all aspects of life. Only then can we release it and begin to move forward.

for and attainment of truth", as the highest principles upon which we base our modern-day society. The United States Constitution begins with "We hold these Truths to be self-evident". Truth has been recognized as the most important principle upon which to base a free and just society. The 2015 Truth and *Reconciliation* Commission and the Elders that helped to shape it from across Turtle Island consistently state that reconciliation can only take place in the face of truth.

Yet in the pursuit of exponential development, growth and wealth, we have ignored this fundamental principle, set aside to win at any cost. The cost, unfortunately, is all of ours to bear. Our current society reflects this lost way—a way built on fear, hate, scarcity: a zero-sum game. This is what the mind of a handful of people looks like whose "dream" this world reflects. And it was allowed to thrive because we ceded to others our own control over our minds and our beings.[29]

As we face a global shift of epic proportions, the real subject is not a health or economic crisis but rather a quest for the future of humanity. Which is the path we will choose for the next millennium of humanity? For a new reality to be possible, the old one must come undone. As we let go of the old, we must focus on building the new. Change is scary, always. But I think we can all agree that the old is not something we want to hang onto. Let us not go back to our old ways of being and doing. Can we take this opportunity to rebuild our world better, on higher principles, and in harmony with life?

My good friend Jamie Miller, a natural engineer and Canada's leading biomimicrist, talks about the adaptive

---

[29] Perhaps we didn't know better before, and this is all part of human evolution to finding the higher way. Perhaps this chaos is necessary to pull out all the trauma to be seen, acknowledged and healed. But when you know better, do better. And I think we can all say that we now know better.

cycle and ecological succession, which provides opportunities for new ecologies to emerge that are more attuned to the new realities.[30] This is our opportunity now.

Rebuilding our societies to last, to continue, to thrive for all, must be based on a foundation of love, in harmony with Great Spirit, with natural law. The most fundamental truth is that we are all interconnected and interdependent, we are all one and love is the fuel of all life in the Universe. In that state of higher understanding, harmony with all life exists. Building a world from this place is our opportunity now. Today's new technology will be outdated tomorrow, but societies based on higher values will never go out of style.

---

30   https://youtu.be/bJxUiwWhBHc

OUR OPPORTUNITY

It is a wild and crazy time to be on this planet—the biggest transition in the history of humanity. It is the reason souls are lining up to be here! It is the greatest show on Earth and it is happening right now, in front of our eyes. We are now in the process of a great conscious evolution, if we so choose it, and it *is* a choice. While the energies (our energies) are rising—and we and the planet are getting flooded with immense light energy, and we are getting a ton of help from the spirit (energy) world—we must *choose* light, *choose* love, *choose* continuity, in order to create the world we all want to be a part of. This is a free will zone, which means we choose our own adventure. We humans have the gift of making a choice to play in the dark or in the light. Making no choice is still a choice, so stand proud and assert your path.

> ## ELDER WHABAGOON BLESSING:
>
> *My Sage medicine burns in the bowl this morning with the smoke carrying my prayers up towards the Creator. Now I wait patiently. The answers will come. My life is made up of all the choices I have made to this day. They may not be everyone else's choices; they may not be another person's "right way". I know one thing for sure though: this is my path, my right way. Peace on Mother Earth.*

The only way to build something new, something better, something that works in harmony with life, is to wean off our dependency on the old systems and those aspects that did not work in harmony with life. In the short term, that means moving to some measure of self-reliance: getting "back to the land"—not to living in the bush, but to building from that way of thinking.

• INVITATION •

> Plant a garden, build a simple greenhouse, put up some solar panels, and dig a cold cellar. Learn to forage again: nature is abundant in food! Let's take back our own sovereignty.

Being independent thinkers and having control over our own lives and minds is the only way we can begin to build the new. And that begins with being able to take care of ourselves. As long as we rely on external systems for all our needs, we will not be able to build something new that we choose.

I grew up camping with my family and in Ukrainian scouts. And while I do not hunt for my own food or purport to be fully self-reliant, the knowledge that I can build a fire, grow my own food, build a shelter, survive in nature, and take care of myself in almost any situation or circumstance has given me a tremendous sense of confidence to be able to make it anywhere I go and weather any storm. This personal empowerment and belief in ourselves are what is most needed now.

Let's open our eyes and wake up to the wonder and brilliance that is each one of us and, by extension, is all life systems. Now is the time to shine our light and be ourselves. We need you! The world needs you! The Earth needs you! Our new collective human story needs you!

Humanity has never built a civilization that has *lasted* because only a system in true harmony with life can continue. Greed, corruption and bad thinking have always led to our demise. And we are now headed for the same game-time decision. We not only have an opportunity to build a *way of being* that lasts; it is imperative for humanity to do so. And it is our collective choice.

Creating for healthy continuity takes a proactive commitment, a personal responsibility as well as a collective one. It takes "working the deal". Benjamin Franklin, after having helped write the U.S. Constitution, came out and said, "We have given you a republic *if you can keep it*". It is not something static that you put on a shelf and forget about. Indigenous cultures were known to have oral traditions because they knew this truth: that, once set in stone, the demise of a society was forthcoming because it becomes too easy to be complacent. Creating for healthy continuity is something we have to *work at* together. And this is a choice—our choice.

LET US ALL CHOOSE TO WAKE UP AND TAKE BACK OUR HUMAN STORY

The only *way* that has continued and continues today is the ancient way of the First Peoples around the world that (contrary to history books, legend and Disney) does continue to exist today. Yes, the Maya are still alive and well, as are the Inca and the Inuit cultures. It is not a thing of storybooks. Many Indigenous philosophies have continued, against all odds, for more than 15,000 years.

What has made these *ways* withstand the test of time? Is it the understanding of our innate interconnection with all of creation, and thereby having respect and reverence for all beings? Is it the knowing that everything is alive and, therefore, we choose to work with nature, not against it? Is it living in community, working together, seeking fulfillment outside of the material realm within the metaphysical (*a.k.a.* our true selves)?

Even with the genocide that has been committed against the First Peoples around the world (and continues today), they remind us of the sacred relationships that exist between one another and with our Mother Earth. And perhaps it is this beautiful *way of thinking* that has had the strength to survive even through the darkest and ugliest times. Because true beauty is power, strength, resilience and love. It is the highest vibrational frequency and power that fuels people to overcome the unimaginable.

Our world has changed and so we must change with it. Yet, if we choose to ignore all the wisdoms of the ancient teachings, we are throwing away the gold upon which we need to build new foundations. How we choose to translate these values and principles of all our ancestors and Elders into a modern thriving economy that works for all, is the challenge of our day.

GROWING UP WITH UKRAINIAN VALUES

I grew up with Ukrainian grandparents who came over to Canada after the Second World War with others from their homeland as "displaced peoples". They had lost their land and fought for their sovereignty, language, culture and way of life under colonial rule. They came to Canada with nothing and started a new life here. Through everything they had endured—losing loved ones, being thrown into a war as kids, surviving, hiding, fighting and then ending up in displaced persons' camps—having to choose their own futures in a land they knew nothing of.

Yet, they never complained nor forgot what life was really about: family, good health, community, humor, sharing stories, helping others, comradery, being on the land, gardening and cooking with fresh food grown by oneself. They were positive and inspirational, uplifting people, grateful for what they had and always ready with a good joke or a smile for a stranger.

My grandfather Ivan tended to his garden, and the love he poured into it reaped incredible harvests every year. He always had a great poem or story to share as we sat down to a family meal. He would play his fiddle, the same one he played with the Indigenous communities he would tell us about, up in Northern Ontario in the work camps he was sent to upon arriving in Canada as a refugee (where he had to work off his boat ride over from Ukraine). He would share fond memories of sitting around the campfire at night with his Indigenous colleagues. Though they did not speak the same language, they nevertheless built friendship through playing music together. His stories were always positive, always in gratitude, and always made us feel the richness in his life, even with seemingly little in the way of material possessions or wealth. He passed away a few years ago at 100 years of age, in good health, with my grandmother and the family at his side, dancing and singing to the very end. He certainly lived a rich and full life.

I was blessed to grow up in community, even as a third generation Ukrainian, and my family fostered these values well. Ukrainian is my first language; I learned to speak English in school at the age of four, which brought some embarrassing stories of trying to find the washroom (sometimes not in time!) or tracking mud through the playroom at nap time. I took Ukrainian dancing, joined Ukrainian scouts (*a.k.a.* nature "boot camp"), and spent a lot of time out on the land.

I grew up with these values and, looking back now, spending summers camping out in the interior of Algonquin Park, canoeing, portaging, sleeping in tents, spending time at our simple cottage on the land, and being out in nature added a tremendous amount of strength and resilience to my ability to be out in the world and do what I have done. This I got from my dad, who always brought us along and taught us how to build fires, canoe, tie a rope, pitch a tent, and use our industriousness to solve any challenge. He has always been a believer in having basic survival skills and fostered in us a joy of being out in the natural world, from an early age. We would carve masks out of wood in the forest and paint them in vibrant colors to reflect what we experienced in nature. Ever since I can remember, he preached sustainability principles, always talking about building off-the-grid housing, well before his time. And he questioned everything, teaching me to think for myself.

My grandfather Walter was an insatiable reader and learning, which has certainly been passed down to me. He wrote a book about his time in war when he fought for Ukrainian sovereignty in the Second World War, and he challenged other authors for taking the humanity out of the experience. He talked about the emotion of being there with people you cared about, the devastating experience of war, and the beauty of the human spirit. We forget the humanity in our stories. Suffering will always be a part of life, but how you choose to perceive that suffering is what builds strength, resilience and grace. He had a plethora of books at his house on every topic possible from history, politics and human rights, to astronomy, meditation and metaphysics, and would challenge us in discussions on complex topics, opening our minds to different ways of thinking and a curiosity about the world.

My grandmother Orysia had a heart of gold and would tell stories of different people and cultures around the world, in a tone saying to us as little kids with blond hair and light skin, that everyone is the same, equal, and part of one human family. Everyone who crossed her path was blessed by her love and kindness—and a few perogies waiting in the wings! I remember picking herbs around the cottage and drying them for teas, which we would drink together as she shared stories with us in the evening. The warmth, comfort and love in her home was something I will always remember and cherish.

Grandmother Olga was very generous with everyone. Everything she had she gave away—from helping to support our education, to sending money and clothing to orphans in Ukraine—all on a shoe-string budget. To her it was not charity; it was the thinking that when you have all you need (and they did not have much by today's standards, but to her it was all they needed: a warm home, a big garden, a root cellar, a happy and healthy family), then you give your excess to those that need it.

She shared her deep love for humanity with everyone she met, was strong as nails, and kept everyone focused on what was truly important in life: family, community, and always a good joke to put a smile on someone's face and make them laugh. She rallied everyone for social justice and could debate any political topic until her last breath at 97 years old, for which—in her ever-determined manner—she waited until the whole family was at her side and each had a chance to say good-bye. Then she waited for us all to start chatting and telling stories, and when she knew we were all okay, she quietly closed her eyes. I could feel her at her funeral whispering into my ear, "Don't be sad! Be happy! Tell a good joke!" She believed in not taking yourself so seriously: if you are healthy, then that's all you need. Get out there and have some fun, and don't forget a big dose of laughter while you are at it!

My mom has been the foundation, the matriarch, the grounding of creativity and love for me. She has always believed in me and that anything was possible, and is always there to help when I need it, no matter how busy she is or the tight timeframe I have (which is usually on 24 hours' notice!). She introduced me to spirituality, to tools that could train my mind to focus and excel, and above all, she fostered in me my own internal genius of creativity and imagination: finger painting, observing nature and reflecting it through art, and being free to express myself (even when, to my parents' shear embarrassment, I went to my first day of high school with sparkles all over my hair!).

I am forever grateful and feel eternally blessed for the abundance of intangible wealth that has been passed down to me. Most of us grew up with a grandparent or aunt or uncle with these traditional values. They may not have had much in the way of material wealth, but they had great abundance of intangible wealth. And that is what makes a life worth living. These are the ways we must all bring back and infuse into a way moving forward. It is not going back to the old ways but bringing the old ways into the new.

BE BOLD WITH YOUR HEART

In order to give birth to a new way, the current systems that do not serve humanity must come undone (as in any growth journey, for those personal growth junkies). Nature does not hold on to what no longer serves it. It lets go and makes room for new, green-shoots to grow. Nature does not resist evolution. Yet we are funny creatures of habit and we hold on to what we are used to, with tight grips—even if it's not working anymore (and really never did)—not realizing that if we were to let go, that space would open up to beautiful new possibilities... always better, always higher consciousness. Do not be afraid of change. Prepare and then allow it to unfold for the highest good of all.[31]

---

[31] Imagine a tree holding on to its leaves in autumn: "No, I refuse to let them fall! They stay with me!" But those leaves strengthen and protect the soil so that the tree can grow even stronger in the Spring.

It is up to us to help others in our human family through this transition with compassion, kindness, love and generosity of spirit. There is a great deal of human suffering in the world today. This is our family and more than that, it is *us*. See a bit of yourself in everyone[32] because we are all part of each other as the one consciousness/energy/I AM. We demand antibiotic- and hormone-free meat, for animals to be treated well, and we cry out for abandoned dogs in the streets in the middle of a cold snap. Yet we have no problems injecting humans with drugs, chemicals and hormones, and leaving many on the streets in the middle of winter. This is our family. We are all one. Perhaps for us it gets too close to home—seeing human suffering that way. But in these difficult times, it is more important than ever to have courage.

INVITATION

Be bold with your heart. For those who have lost hope and vision for a better tomorrow, remind them that we are building an incredible world. Remind them to have hope, to believe, and that it is possible and not as daunting as it may seem. Together we can do anything! Tell them what your vision looks like, what it *feels* like. Remind one another that we are the co-creators of our reality and that we can choose to take control of our minds again. Show others that we are powerful light beings by shining *your* light bright.

---

[32] The Transcendors, Spiritual Teachings by Rik Thurston. A group of enlightened souls that assist in humanity's journey and spoke about the path towards continuity.

And for those staunch consumers out there, no, it is not a choice of giving up everything to be on the "sustainability" bus. The economy can be 1000 times bigger if created on foundations of love[33] and not fear. Fear is a scarcity mentality; it is a zero-sum game: "If you have more, I have less". Love is an abundance mentality, meaning *infinite* abundance if we create from that mindset. So, have a little trust and faith!

Many shamans and teachers say that it will only take 10 million people to choose continuity for humanity, for us to get on the side of renewal. And people are waking up quickly! This means starting to think and act as the ONE I AM[34] that we are. We have been sold a bad system and we bought into it for too long. It doesn't work. And we all know it! But we are stuck with it unless we can create something better. So, what are we going to create? It is up to us to be the world that we want to create and begin to live our true potential.

OKAY, STOP RIGHT THERE!

Before you go and say that we are past the point of no return, that it's too late, that it's just not possible—overpopulation, food and water scarcity, pandemics, financial systems in collapse, draughts, climate destruction,[35] etc., etc., etc.—stop! Everything is energy and your

---

33 *A.k.a.*: creativity, imagination, passion, beauty, inspiration, ingenuity, genius.

34 The One "I AM" is all there is—it is us, it is Source, it is all life, it is all of existence. It is the Source in each of us.

35 I prefer "climate destruction" to "climate change". It is far more accurate. The climate is always changing; that is its *nature*. Seasons change, as do climate cycles. Plus, we are all rising in consciousness, including Mother Earth! The term "climate change" creates divisiveness, not unity, because it is too easy to refute—of course it is *changing*, nothing ever stays the same. Climate destruction is when we know better and still choose the harmful path.

mind is a powerful co-creator of your reality. So, over time, what you think and put your focus on is what you create. Is doom and gloom really what you want to stand behind as your contribution to the world? A system perpetuated by some pretty limited thinking? Do you want to be *that* guy or *that* girl?

I am not suggesting that you ignore 'reality' as you are currently experiencing it. Be aware, keep your eyes open, and stand up for what you believe in. Mother Theresa said she would never go to an anti-war rally, only a peace protest. Focusing "against something" is still focusing energy on that thing you *don't* want. That is manifesting 101! Let us focus our mind and energy on what we want to create and be a part of. Dr. Joe Dispenza, the well-known neuroscientist, talks about this when he teaches how to re-program your mind to create the reality you wish to experience.

You are right: it's a shit storm out there that we have created. It will take a collective mind thinking from a wholeness perspective[36] to create a collective future of well-being. So, we need everyone to start putting our mental power to something more in harmony! Together it is possible to create that world. It will take collective intention and collective action. Start building greenhouses, a garden, help a neighbor in trouble, smile at a stranger. We can make this transition painful or powerful. Consciousness is not a final destination, rather a process of how you choose to live your life. We have a great deal of help from the Universe at this time. But we must first choose to help ourselves. Then the miracles will happen.

When the COVID-19 pandemic started, I went to the largest grocery store in downtown Toronto for a few things. I was not expecting what I saw. I walked in and the entire store was completely empty. All the fresh fruits

---

36 Wholeness is the state of fully being in unity with ourselves and life, which equals complete health and harmony.

and veggies, meats, fish, frozen goods, non-perishables—everything gone. The image of the magnitude of such a big store completely empty was shocking. I was stopped in my tracks, in shock.

I took a second to look at it and see it for what it was.[37] Then I took a deep breath, centered myself, got a shopping cart, and calmly and confidently (standing in my connection to Source) I started walking around, breaking through my initial fear. I felt a sudden feeling of peace wash over me and I smiled. I found my way to my calm center, my true self, and that allowed me to get out of my fear-mind into my heart-center, my *humanity*, and see from a different perspective. I began to chat with the staff, thanking them for their tireless work stocking shelves around the clock. I began to feel so joyful that I started singing along to the music playing, with a smile on my face. All of a sudden, the exact food I was looking for would appear in an aisle I was walking down. And then the next one and the one after that. I ended up with a cart full of what I came for— all healthy organic food.

I knew I had received blessings because I was putting *out* the blessings. Keep moving towards love and light. Don't get caught up in fear. It will hold you back. There is a beautiful future that awaits those that get to it and build it. Anything is possible in that reality. Let us all start now!

OF COURSE IT'S POSSIBLE!

Yes, OF COURSE we can create a world where everyone has an abundance of nutritious, life-enriching food, clean water, free clean energy, a safe and healthy place to live, good health and well-being. We can create a world where

---

[37] We must not turn away from what we are afraid of or what pains us (unless we are in imminent danger). The only way that we can dissipate the dark is by shining the light on it. That means looking directly at it, sitting in those uncomfortable feelings, and breathing light into it.

the oceans and lakes are not only clean but are alive and cleanse our bodies and our souls when we go into them; an Earth that is strong and works with us (when we work with her) to nourish and provide for all of Earth's inhabitants; air that fills our lungs with beautiful oxygen that allows us to soar in life.

Yes, OF COURSE we can create a world that is peaceful and prosperous for all, filled with people happy and healthy. Where everyone is living their true selves and thriving. Where communities are vibrant and dynamic. Where creativity and ingenuity rule the day. Where everyone is contributing and valued for their gifts. Where we smile and look forward to our day, and the world smiles back at us. Where we are in a constant state of flow and manifest our heart's desires. The world around us reflects our inner state. When we learn to create from the inside out—from our heart-center, from our true selves, we create a reality that SPARKLES. Use your imagination and imagine the world *you* want to create.

We can create anything we want![38] Hello?! We are the co-creators of our reality! Soooo, what are we going to create together? Look outside—is that really the best we've got? What does that better world feel like to you? Look like? Taste like? Smell like?

Why create from the heart-center? Well, our heart-center is *already* tuned to harmony, plugged into the "motherboard", a gateway to higher realms of understanding and being. So, when we create from our heart-center, we create from our true essence. And then, what we create is automatically in harmony with all things, and most of all, it is tuned to our true purpose.

Yes! Stop the search party! It's been inside us all along! No need to stress or over-think it. Get into the rhythm of your heart and start to flow with the natural intelligence

---

38 As long as it is within the bounds of natural law, which is a good thing! More on this later...

of the Universe that we are all a part of. Take off the layers to reveal your true genius that's been hiding there all along. We all need *your* genius to make this world thrive!

The more you live there, the better your life will be and the better our world will be because we are all interconnected and each of us is contributing to creating the whole human story together. Right now, we are choosing to create a world based in fear, hate and scarcity by allowing others to create our reality *for* us, by telling us how to think, act, be. I think we can do better. And not just *better*... WAYYYYY better!

I am challenging us all right now to take back our human story and co-create A VISION for a world that works for all, and that continues, for all life! It is our CHOICE what reality WE accept. Do we CHOOSE a fear-based reality of scarcity, loneliness and suffering or a HEART-CENTERED reality—one that is built on the energy of love, abundance, health and happiness? I think that would be AMAZING! I CHOOSE heart-centered all the way!

It is not going to be easy. It requires something really tough... for each one of us to *be our true self*. To go CRAZY being *you*.[39] And not the you that your lower mind is telling you to be,[40] but that quiet voice deep inside that is crying to get out. That place where your creativity, connection and radiance come from. It is a journey of choosing full acceptance of who we truly are. We each have individual uniqueness and gifts that no one else has. If we can all manage to get back *there*, what a spectacular world it will be.

---

39 Guru Singh, one of the great Kundalini Yoga master teachers and rock stars of inspiration, always encourages people to GO CRAZY BEING YOU!

40 No, your true self doesn't look like your friend or the latest social media celeb. Your true self has its own opinion, own voice, own smile, own experiences and own way to shine your light in the world. *That's* the one we want!

HOW DO WE DO IT, YOU ASK?

We take the leap first by making that CHOICE, setting that *intention* and making it as strong as possible, then taking simple steps in that direction. Align your compass, your GPS, and then start the journey, one step at a time. If you make a wrong turn, no problem. Just like Siri, when you set your GPS to your final destination, it autocorrects and you get brought back to the right path. The next ten years will determine the direction that our planet will take—our future—and it is more important now than ever to make sure, first and foremost, that our compass is set to the final destination that we choose. There may be darkness before dawn, there may be challenges and obstacles along the way.[41] If you set your final destination and continue to live by your heart-center, you will always be brought back and up into the light.

An Ojibwe Elder[42] shared a teaching at a sweat-lodge ceremony I attended at his house a few years ago, together with members of our organization and Indigenous leaders from South America. It was one of the most beautiful and humbling experiences I have ever been a part of, for the energy in that lodge was incredibly powerful. Peter shared that the Ojibwe First Peoples believe there are branches of the tree that take you onto smaller learning journeys along the way, but that you will always go back to the tree trunk and grow up towards the sky.

Your higher self knows your true path and knows exactly how to get there. But it won't interfere until we say so (a thing called "free will" and "choice"). Oops, did we forget to choose?!

---

[41] Just like any good hero's journey.

[42] Elder Peter Schuler, of the Mississaugas of the Credit First Nation.

• INVITATION •

Stop right now everything you are doing and close your eyes, put your hands over your heart, and take three deep breaths. Then choose a world that is thriving. Choose love. Choose peace. Choose compassion and kindness. Choose yourself. Even if that is not the reality you currently know or experience, *just saying those words* infuses you with that energy and begins to take you there. Your higher self knows the direction. Don't worry about figuring out the details of where that is or what it looks like. Right now, just make the conscious choice and set the intention. Period.

Good? Easy? Okay, now that your GPS is activated and knows where it is going, fasten your seatbelts and let's go!

CENTERS OF LIGHT

Some time ago, in a meditation, I had a vision of many Centers of Light: centers of extremely high-vibrational energies, love and harmony functioning in this new reality. They were all interconnected with beautiful strands of light, almost floating in space amidst darkness.

We may not be able to lift and transition the whole planet tomorrow, but we can create Centers of Light wherever you are that work to uplift and bring into the light all those around them. We can connect these Centers—be they communities, neighborhoods, groups of people, or just one individual. These are places where people are choosing to live in the new way. It could also be just one person—just you—who is choosing a higher path and radiates that high-energy frequency to all with whom they come in contact.

    In my vision, these Centers multiply and then together lift up the planet. This is the potential we hold today.

WE ARE ENERGY BEINGS (*A.K.A.* SUPERHEROES)

"If you want to find the secrets of the Universe, think in terms of energy, frequency and vibration. The day science begins to study non-physical phenomena, it will make more progress in one decade than in all the previous centuries of its existence".

—Nikola Tesla

We are all energy.
We are all interconnected.
We are all one.
And love is the fuel of the Universe.

Me: *What?*

Pigeon:[43] *This is the basic premise of natural law.*

Me: *Huh? What is natural law?*

Pigeon: *Well, all life in the Universe is subject to natural law. It defines our rights and responsibilities to each other and all beings.*

Me: *So, like, rules?*

Pigeon: *Sort of, but more like the natural rhythms of things. We are all subject to them. It is the "mechanics" of the Universe and all things in it, much like seasons or the tide of the ocean. It just is. When we are in the flow of the natural rhythms of life energies, we are in harmony with life. When we are against the flow, we are not in harmony. Simple as that. And it's infused into all things, so we don't have to worry about it. Just tune in with the natural rhythms, and you will be in the flow of life again!*

Me: *Okay, so what's the natural rhythm?*

Pigeon: *The essence is healthy continuity. All beings are always striving to continue living, in as healthy a state as possible—their best self. The whole understands how to make this so.*

Me: *So, like, Seven Generations?*

Pigeon: *And beyond.*

Me: *It's different than 'sustainability'?*

Pigeon: *Yep. Sustainability is sustaining something in a static state. Our planet (which you humans are a part of) is dynamic and ever-changing yet striving to always be*

---

43   a.k.a. Great Spirit, the I AM, the motherboard. Wearing the appearance of a pigeon, out on my morning stroll.

in balance (other than you humans, of course) with all things for continuity of the whole of existence. If the whole thrives, we all thrive. It's about applying these understandings in all we do. So that things last. Not in a static state, but in an ever-changing harmonious symphony.

Me: *I see...*[44]

OKAY, LET'S GET METAPHYSICAL WITH IT!

Everything is energy and is part of the First Source energy—the ONE I AM.

The First Source energy, which we are all a part of, created this beautiful playground called Earth. It then created sub-sects of itself in order to experience itself and its creation in its many wonderful layers, colors, aspects and themes. It is the beauty in diversity that we all came to experience. We are all parts of that one energy or consciousness. The First Source energy is in all of us, infused through all of existence. We each have the "God Spark" in us, as they say.

Imagine a great ocean. All the drops of water are individual drops yet are also part of the whole. Each drop of water holds the essence of the whole—it holds the molecules of an ocean, the salt, and the minerals. It knows it is part of the ocean, and without all those drops of water, there would be no ocean. We are like the drops of water that are parts of the whole of the ocean. But we can't truly ever be separate from the whole because our energies are

---

[44] Yep, this was the conversation between a bright-eyed pigeon on the seawall and me in Vancouver's Stanley Park one rainy afternoon as I went for a stroll. Twenty minutes of standing in the rain talking to a pigeon definitely did get some stares, but messages of genius come in many forms, so you gotta take them when you can get them! Einstein, Tesla and Mozart all claim to have received their "genius" when out in nature, amongst the trees and birds, in meditative states of being. So, get out there and engage with the Everywhere Spirit in its many forms!

interconnected and can never, will never, exist outside of the one whole of consciousness.

We are energy beings having a human (physical) experience. A long time ago, we forgot this and got stuck in, attached to, physical *form*, a denser state of being. We were once able to jump in and out of form, simply coming here to experience the wonder of the oceans, the forests and all our relatives, in material manifestation. Many shamans still practice this today, what we know as shapeshifting. Their spirit can disattach from their physical form and experience other forms, as we all were once able to do. When we became attached to form, we began to disconnect from one another and from the First Creator inside and around all of us, because we forgot our connection with all life.

But this physical form is not our true essence, and it is time to wake up and remember. Our true essence is a great connection with all that is. This means that we are never alone. When you feel alone, connect to the multiverse within you and all around you. Know that you are a deep part of all of Creation. Have a conversation with one of your relatives—the bird on the fence, the ant crawling across the front lawn, even the lamp on your desk... all is consciousness. Everything is consciousness and we are a part of everything. Become aware and acknowledge all the different aspects of our conscious experience. It is all here to teach us.

ELDER WHABAGOON BLESSING:

Lessons are hidden under every leaf and rock. Look up! and the winged-ones that soar will teach you how to fly.

The First Peoples believe that everything has spirit because everything *is* spirit—the land, the water, the rock and the tree. The land is sacred because we are sacred. We are all from one Source consciousness and together make up that one Source consciousness. As such, we are not separate from nature. We *are* nature. We are the river that runs through the forest; we are the trees; we are the rocks; we are the oceans; we are the eagles and the stars in the night sky. How can that be? Well, because we are *we* and we are *one*, all at the same time—one great big ginormous energy field!

Here is a great song to listen to or, even better, to chant! *Be* the sound and feel your body vibrate!

WE ARE WE AND WE ARE ONE. WE ARE WE AND WE ARE ONE. WE ARE WE AND WE ARE ONE. WE ARE WE AND WE ARE ONE. WE ARE WE AND WE ARE ONE. WE ARE WE AND WE ARE ONE.[45]

"Can anyone doubt to-day that all the millions of individuals and all the innumerable types and characters constitute an entity, a unit? Though free to think and act, we are held together, like the stars in the firmament, with ties inseparable. These ties cannot be seen, but we can feel them. I cut myself in the finger, and it pains me: this finger is a part of me. I see a friend hurt,

---

45 Hummee Hum Brum Hum—a high-vibrational sound and combination of letters in Sanskrit that create a resonance which raises your frequency (*a.k.a.* tune you up), scientifically measured. [James, Kevin. "Humee Hum Brahm Hum". *Life is Perfect*. Kevin James, 2013. CD.]

and it hurts me, too: my friend and I are one. And now I see stricken down an enemy, a lump of matter which, of all the lumps of matter in the Universe, I care least for, and it still grieves me. Does this not prove that each of us is only part of a whole?"

—NIKOLA TESLA

ALL LIFE IN THE UNIVERSE IS SUBJECT TO NATURAL LAW

No one is outside of it. Phewf! That's a relief! We all belong to each other and to our natural environment. That also means there are some big rules that govern the Universe that we don't have to worry about—seasons changing, birth and death, decay, renewal. We are part of this cycle. We forgot that we are a part of the Earth and part of natural law. We can speak the language of nature and we can flow with the rhythms of nature again. We have just closed our minds to our natural world, to our true selves, and the understandings that come from remembering our non-separation.

THE ESSENCE OF NATURAL LAW IS HEALTHY CONTINUITY

Continuity is building to not only last but to exist in harmonious balance with all of creation now and in the future. Life is a great big, beautiful symphony! When a forest thrives, all life forms in that forest thrive as well. When the forest is sick, all the life forms also become sick. This is what we are experiencing today on a global scale.

Continuity is a way of thinking. It means a consideration of the whole, building for harmony now and for the future—TODAY. We build structures that may last a few decades at most and do not consider all other beings (including most other *human* beings) in the process. Our institutions rot from the inside out. Our resources get depleted and our natural (clean) food and water systems are toxic. Because we don't build for continuity.

It takes a complex understanding of the aspects of our interconnected whole, how they work together, and the impacts of our actions, our *energy*, on others—a never-ending feedback loop that reflects the interconnectedness of the whole. It is not a static state; it is a natural, dynamic and ever-changing dance. Do you want to be frozen in time without changing, growing, shifting and evolving? There is a natural rhythm to life, to all things. There are cycles and shifts. Continuity is built from a place of natural law, to continue in the flow of natural law.

It also does not mean controlled. It does not mean a river that has water but that is no longer truly *alive*—that no longer has life or life-force[46]. We know when we are being controlled and when we are allowed to flow naturally. Just because we can't see it, doesn't mean it isn't real. There are trillions of microbes in the soil and the waterbed that are also part of creation and therefore part of us. Science has yet to be able to come close to explaining the magnitude of interconnectedness that makes up our existence in this physical and non-physical form.

The First Peoples live by the laws of continuity—for seven generations and beyond. In the Indigenous worldview, everything is to be considered today from the perspective of all life forms, for seven generations out. There is no word for "sustainability" in any traditional Indigenous language that I have come across. It is not a separate distinct consideration. It is simply a way of thinking that considers the whole in all we do. Now this is complex stuff!

---

46   Where our true nourishment comes from—it is actually from life-force energy. That is why we can eat and eat and never quite feel fully nourished because much of our food, water and subsistence today lacks life-force energy. The energy has become weakened. There is nothing like fresh veggies from a garden or wild meat. You can feel the high-energy that still resonates there and that we ingest.

Spending time with Jerry on the land in his territory was the most incredible experience. The land knows Jerry and Jerry knows the land. They are one and the interconnectedness of all life and spirit that resides there is understood, even when not spoken. Jerry points out where he hunted for moose and where the moose calve, where the salmon spawn, and the grizzlies den. There is deep respect there in how he speaks of the animals and the life forms. He talks of how the interconnected ecosystem always strives to maintain balance, pointing out 50 variables that take place from simply seeing the yellow flowers start to bloom on the mountainside. Jerry still remembers the language of nature and *sees* the complexity of the systems, as well as the solutions that a deep relationship with the natural world provides. He often sends me natural medicines from the land for a healthy liver, cleaning the blood or any number of important functions. He tells me that he learned about the medicines from the moose. When they would hunt for moose and their livers were grey, they knew they were sick. They would watch the moose and the moose would find medicines to heal their livers and the next hunt the livers were healthy again. Through observation, over 15,000 years, there is a rich understanding of complex systems and a way of thinking that sees more, *values* more.

This kind of knowledge comes from a deep understanding of the land where you are from. And it is this deep understanding of the interconnectedness of life that is necessary in order to make decisions today for healthy continuity tomorrow. There is no way we could readily know from a textbook all the infinite lifeforms and relationships that take place in a constantly changing symphony. That is why it is said that historically the Indigenous didn't ask *why* it was the way it was: rather, they accepted that it *is* (as Great Spirit's creation) and sought to understand *how* they could live in a greater state of harmony and balance.[47]

---

47   The Transcendors, Spiritual teachings by Rik Thurston.

The Nuu-cha-nulth Indigenous culture has as its highest law a phrase called IISAAK[48] which means to observe, appreciate and then act accordingly. This is how we learn to make our own decisions from a place of connection and harmony that is of service to life. And when we create from our high mind, we are automatically tapped into the one Universe that is everything, and therefore can immediately tune in to, feel and understand all life systems around us. This is how we begin to *see* more. And this is how we may build for healthy continuity.

Frankly speaking, our society is built on limited thinking because it is not built to continue, let alone continue in harmony with, well, anything! In 100 years, what would this planet and our human species look like if we stayed on this course? Would we humans make it to 100 years? 50? Hmmm. Either we learn to thrive *together*, or we destroy our human forms *together*. Remember, we are all in this together. The laws of nature do not make exceptions if you have lots of money, a big house, a fancy car, or a private jet.

Robin Wall Kimmerer in Braiding Sweetgrass puts it beautifully:

> *If one tree fruits, they all fruit—there are no soloists. Not one tree in a grove, but the whole grove; not one grove in a forest, but every grove; all across the country and all across the state. The trees act not as individuals, but somehow as a collective. Exactly how they do this, we don't yet know. But what we see is the power of unity. What happens to one happens to us all. We can starve together or feast together.*

---

48 Eli Enns, of the Tla-o-qiu-aht First Nation and co-founder and President of the IISAAK OLAM Foundation, that works to establish tribal parks (Indigenous conserved areas) across Canada.

MATERIALISM IS SO LAST YEAR!

Don't be afraid to give up your attachment to the material realm. This is not to say do with less, it just means to not get *attached* to the material world. Let go of what holds you down and get out there and have fun! Quan Yin[49] reminds us that material things are part of the Earth vibration and are only on 'loan' for us humans to use. Enjoy this experience in this incredible place we have come to explore. Remember, we are all stewards of this Earth and what we have in the material sense is only for us to learn how to truly be custodians and explorers of this Earthly realm. It is attachment to the material that causes the greatest suffering. When we let go, we realize that abundance is in fact everywhere.

NATURE'S INTELLIGENCE

Nature builds for continuity. A tree will not grow fifty feet taller than the rest in the forest because it inherently knows that if the other trees do not survive, it will not survive.[50] Trees understand how to minimize risk, in order to maximize a thriving, living ecosystem. The more diverse, the healthier the ecosystem. How did we forget this? It is the intelligence of nature not to destroy itself but rather sustain itself in a collective cycle of life. It is complex engineering because it considers all things. Imagine the trillions of variables that are part of the complex web of life that is nature. Now that is incredible engineering!

   Nature is built to last. The cause and effect towards continuity is built into the design! And we are a part of that complex design, because we are nature. Hello?! Yet we waste so much time, energy and resources cleaning up our messes because we just didn't do it right from the

---

49   Quan Yin, as taught by Shin Yin.
50   Jamie Miller, Biomimicry Frontiers.

beginning. Let's take a moment to stop, observe and apply these teachings to how we build our world.

When you breathe out, a tree breathes in. When it breathes out, we breathe in. Isn't that incredible? It just happens, all day, every day, all around the world. If life dies, we die, and we are slowly dying (cancer, stress, dis-*ease*). Where do we think we get our life from? Last I checked, we don't buy our air on Amazon![51] But it is not about being an environmentalist or stopping the consumption train. And it's not about limiting yourself for the whole. In fact, in this way, there is so much more.

> ### ELDER WHABAGOON BLESSING:
>
> *Respect our trees. Love our trees. Our trees are beautiful. Honor them. Sit with them. Talk with them. They will hold your words dear. They provide life. They exhale what we inhale—they give us life. They have eco-systems underground and a support system that is very family oriented. Big trees, little trees, bush or hedge talk with them. Hug a tree and most of all say miigwetch to our trees.*

When the world around you thrives, you thrive. Why? Because we are all one! It is about being aware of all life and our role within that whole. It is about the *how*. How do

---

51   Do not be fooled by the clever name 'Amazon'. It does not mean the actual forest that has been coined 'the lungs of the Earth'. No amount of 'Amazon' deliveries will bring you life like the vast forests that release oxygen and store carbon dioxide so generously to create a climate that humans can exist within on this Earth, *by design*.

we relate, interact, reciprocate, consider, acknowledge? How can we pause and consider more? How can we foster greater diversity? How can we live in better relationship with each other and the Earth?

It is not about preserving and protecting, sustaining in static states. These are fear-based energies. Nature plays, nature interacts, nature *relates*. Know the limits of natural law, respect Mother Earth, treat all things as you wish to be treated, and then get out there and PLAY with the energies! Engage, interact, frolic, prance and be merry! Roll up your sleeves and get your hands in the dirt! That is the game. It is a game of joyously experiencing the energies around us in a multitude of forms.

My dear friend Arzu, a powerful healer and herbalist, often shares herbs with me for various things (full moon harvested and wild crafted of course!). They are full of energy and take on the darkest critters that may have come into my energy body. I remember the first time she shared some herbal tinctures with me for an upcoming trip to the Amazon that was 3 months away. There were no instructions on the bottles she handed me. I looked puzzled and asked how much I should take and how often. She looked at me, with this big grin and a glimmer in her eye, and she said, "Girl, how am I supposed to know what your body will need in three months from now and how often? You've got to connect with your body and listen to its intelligence! Your body knows. Connect with the plant and let it heal you". We are so used to being told what to do and how, given "one size fits all" prescriptions, that we forgot about the intelligence built into us, that is in nature and in all life. That was a lesson in trusting myself again.

INVITATION

We are on the most diverse planet in the Universe—it is literally a biodiversity library that souls are lining up to get onto and experience. Play with the I AM that is water. Breathe in with intention the I AM that is air. Sing with the I AM that is a robin. Stomp your bare feet on the Earth. Reach your hands to the sun and thank him for the beautiful life-giving energy. Smile, laugh, and become an explorer of life again!

AI VS. HI

Technology can be a wonderful thing, when it is programmed, intended and used to support our vast innate human potential, in healthy ways that foster life. It is certainly helpful when used to unite human ingenuity and creativity to build a better world, to calm one another in chaos, and to spread love and kindness around the globe. But the idea is ludicrous that we look to AI to go "where no man has gone before" because we believe we have reached the peak of what human intelligence is capable of.

The reality is that we have not even scratched the surface of our Human Intelligence (HI) or Human Technology. When you start to realize that you had a hand in creating vast oceans, that support trillions of microbes and other life forms; the forests and the lakes; the animals and the birds; complex ecosystems and forms, we will start to uncover our true *limitless* potential—to heal, to communicate with each other and the natural world without words, and to manifest to our heart's content. So, before we give our power away to robots, let's first uncover the vastness of our own true potential—our own limitless internal technology.

When we unlock our creative potential by remembering who we are and by tapping into the real motherboard, we will literally be capable of anything. We can, when we train our minds, create into physical form anything with the power of our minds and our hearts—without digging up resources, processing them and transporting them across the world using fossil fuels. That is our potential when we develop our true internal superpowers.

WHAT IF YOU COULD BE A JAGUAR FOR THE DAY?

What if you could fuel your car just with your belief? What if you could uncover the secrets of water or plants or trees by simply having a conversation with them? I had the incredible experience of popping into a tall beautiful plant[52] in the Amazon while sitting and meditating by the river[53]. I could feel the sway in the wind, the sun beaming down on my soft leaves and the nourishing water that I was soaking up from the soil.

We have been playing in one room of the house and think we have mastered the house, when in fact there are many other rooms we haven't yet stepped into yet. Don't be afraid to go exploring beyond what we are told we can play in. We have the whole Universe within ourselves! It is so incredibly vast, beautiful, mind-blowingly cool and awesome!

What will you be? What will you create?

---

[52] It is said that plants carry divine feminine energy and trees carry the divine masculine, so it is easier at first for women to connect with plants and flowers, while men with trees, but we all carry the masculine and feminine so it all works!

[53] No, not as part of any plant medicine experience. Simply by connecting to the 'motherboard'.

PROGRAMMING THE NEW HUMAN STORY

Technology is the infrastructure that drives our society. It can be programmed to support thriving systems or to inflict harm, and all sorts of ways in between. It all comes down to intention. When you go into a new business or start-up, see if you can feel the intention behind the business. Was it just for personal gain or was it to add value, serve society and then personal gain? We can always feel the passion that drives a truly successful venture that came from a place of pure creativity and ingenuity.

Intention is the energetic imprint that we put on something and it flows through into our creation. It is the source of our manifestation. The way we think is the way we create. Technology cannot operate without our intention. Let's begin to build technology (software, games, communication towers and computers) *programmed* with positive intentions, for a healthy, thriving world!

WE ARE IN THE GREATEST VIRTUAL REALITY GAME IN THE UNIVERSE RIGHT NOW!

Did we forget the greatest game on Earth—a virtual reality experience like no other, that we are literally creating with our minds every second of every day?! Talk about "high-tech"—we don't even need expensive machinery or equipment. Our minds are that advanced! We are co-creating each step we take through how we choose to react and respond to situations in our lives.

As energy beings, this game is multi-dimensional, so not only are we choosing our actions, we are sending out frequencies that can either build people up or down; thinking thoughts that can either make us strong and healthy, or sick and weak; open to abundance, or closed to scarcity. It is the greatest game on Earth and it is ours to experience and add to! The goal? Make it better, if you can. Bring as much joy, happiness, laughter, comfort to as many people as you can. Create well-being for as many people as you can. It's not *what* you do but *how* you do it that counts. It is a game of service, of human relations. Take steps towards building a society that withstands the test of time, that lasts. That is the real challenge of our time.

It's the most complex and exciting virtual reality game that can ever exist, and we are in it right now, all of us together. This game has 7 billion players, playing all at once. All interconnected by a web of energy that allows us to communicate in multi-sensory perception and to *collectively* intend to take action. Do we all want to collectively intend to clean the ocean, rebuild our food systems, create dynamic communities? The brilliance of this game is that we can collectively intend anything, and it will be so if we can get enough of the players to intend with us. It won't take 7 billion players, maybe just a few million. That's more than some people have on their Instagram following! Imagine if we all focused energy towards clean oceans, nutritious food for all, love for all children? Intention is energy and focused energy is power.

We are entering the age of INSTANT KARMA, which means real time cause and effect experience. Now that is cool! We are choosing our own adventure and we are now choosing to be able to experience each decision and choice we make in this game, *in real time*. You will start to notice that what you put out there you attract back right away. We chose this upgrade in our software because we all wanted to evolve in this lifetime. That means we have to learn how to make better decisions and fast. The brilliance of this upgrade is that we now get to experience our choices as we make them so we can auto correct and choose better—for the sake of humanity, our planet and ourselves.

To truly be a wiz and master the game, we must begin to make decisions from the heart-center. That means waking up and remembering who we truly are and living from our heart-center. We do not need to go through an intermediary to tap into our connection with Source, God, the Universe, the Divine. It is always inside of us and we can tap into Great Spirit, anytime. Our higher self is where our true intelligence lives. Our low mind is great at memorizing data and dispersing knowledge. It cannot drive our life's purpose. That comes from our creative mind, our higher intelligence, and that is guided by our heart-center.

When we remember that we are limitless beings, living a temporary existence in this human form, we will break through the self-doubt and limiting beliefs that hold us back from our true potential. We are vast and our consciousness infinite. Only we can play this game with free will. How many will choose to play the new game?

WE ARE LIMITLESS! WE ARE LIMITLESS! WE ARE LIMITLESS! WE ARE LIMITLESS! WE ARE LIMITLESS! WE ARE LIMITLESS! WE ARE LIMITLESS! WE ARE LIMITLESS! WE ARE LIMITLESS! WE ARE LIMITLESS! WE ARE LIMITLESS! WE ARE LIMITLESS! WE ARE LIMITLESS! WE ARE LIMITLESS! WE ARE LIMITLESS!

WE ARE ALL INFINITE CO-CREATORS

We already know that we are co-creators with Source Energy that created all that we are, touch, see and experience. While the Everywhere Spirit created the canvas, we have been allowed to choose our own adventure. And we can choose to collectively create our story. In order to truly respect that incredible gift—the miracle of being alive, the incredible experience of being part of this infinite learning library, the life forms, the creativity, the wonder, the beauty, the awe, we should really try and put our best foot forward, wouldn't you agree? We should strive to create that new story from our best selves, our highest understandings, our more creative endeavors—from our high mind, our limitless mind. This is the mind that is tuned into Source Energy—that invented water and air and the complex systems all around and within us. Isn't it just incredible that we were a part of that?! And wouldn't it just be incredible to see what else we could create from *that* place?

Wow, so cool! There really are no limits! So, let's blow open our imaginations and get to work drumming up some inspiration for our new collective story!

My good friend Becky Big Canoe,[54] a brilliant, wise and incredibly strong Indigenous Grandmother, designed a house that is in true harmony with life. I was so inspired when I saw it that I have committed to ensuring that one day it is built, to show the world what is truly possible with great creativity, multi-dimensional thinking and a deep caring for the wellbeing of people and planet. The house literally hugs you when you are in it. The walls breathe and soothe you in their natural design and materials, decreasing stress and supporting health; it considers the natural elements and is optimized for energy efficiency with the sun and the wind, rainwater collection and compostable toilets; it has a true respect for the land upon which it is built, working with the land and enhancing life systems; it incorporates food production and food preparation for nutrition and wellness. It also nurtures the people in it with provision for an Elder in the center to share wisdom and stories with the youth, passing on knowledge and teaching the next generation. It is a beautiful creation. What will you create?

In order to create, we must be able to imagine what we want to create, to feel it, taste it, smell it, hear it, and hold that vision. But more than anything, we must begin to *believe* again.

I attended in the Berkshires one summer a beautiful kundalini yoga meditation and music festival called Sat Nam Fest. It was a week of intense meditations and exercises to train the mind and break through habits and resistance, mixed in with beautiful heart-centered music, dancing, singing and conversations, set in nature. After a powerful two hour meditation set with a renowned guru from India, I literally felt the entire Universe within me.

---

54 Of the Chippewas of Georgina Island First Nation and Founder of EnviroNative Training Initiatives.

When we were finished, I got up to go use the port-o-potty outside the big tent in which we had been practicing, along with many others.

Just as we went outside, we noticed that the dark clouds had set in and had completely covered the sun. It looked like a bad storm was imminent. Feeling incredibly connected with all of creation in that moment, I looked up at the clouds, raised my arms to the sky, and in a loud and confident manner, said "May the clouds part and allow the sun to shine through!" I waved my arms in a parting fashion as I said this, with a big smile on my face, imagining what the sun would feel like, its warmth upon my face. Immediately, as if on my command, the dark clouds completely parted and the sun shone through brightly, right where we were standing. I was somewhat stunned for a moment, as were the others in line with me. Then we all just smiled, knowing that in fact we are all connected with all of life and that we are literally capable of anything that our minds will allow us to believe is possible. In that moment, I could feel the Universe winking at me again, as if I had been shown a secret.

WE CAN ONLY CREATE IN THE PRESENT MOMENT

Only here and now. Staying in the present, observing what you see, what you hear, what you smell, what you taste and what you sense (intuitively) is how we begin to train our minds and our senses to be multi-dimensional, powerful, co-creators again.[55] From here we can truly observe and begin to consider so much more. We can only tap into our collective motherboard from the present moment—truly tapped into ourselves.

---

55 Manari Ushigua, a powerful shaman and good friend from the Sápara Nation in Ecuador's Amazon, talks about the new education for humanity being to remember and re-train all our senses to be fine-tuned again, allowing us to begin to open our blinders up to seeing and experiencing more.

> ### ELDER WHABAGOON BLESSING:
>
> *What are you grateful for today? Be mindful today and stay in the moment. Take a moment to gently close your eyes, inhale deeply, and exhale long and slow. Clear your mind about today, tomorrow, and remove the clutter. Think about yourself in this very moment. What are you grateful for?*

In that place of true connection with self, there is perfect equilibrium. Our emotions are in harmony. It is a state of being—of being fully alive and aware and feeling *connected*. It is a place in our physical bodies that I equate with my solar plexus. It puts me into my body, my power center. Where is your power center? What does it feel like when you connect with it?

Walking through the Amazon, I was taught to be so present that it would be possible to identify every sound, smell and shadow. Of course, I was not able to—who are we kidding!—there must have been thousands of sounds and smells! Far too many for my urban mind to distinguish. But my friend Cawetipe from the Huaorani tribe could. This is for survival because the First Peoples community I was visiting needed to know growing up in the jungle what was around them so as not to be attacked by a puma or jaguar or bitten by a poisonous snake. This necessity creates a sense of total presence and stillness inside.

In our world of conveniences and separation from nature, we no longer need to be aware of our senses as we once did. So, we spend most of our time in the past or in the future (*a.k.a.* our minds), worrying. And I was the queen of worry! It would make me stressed out and sick all the time.

But we are now going back to needing to retrain ourselves to be there, not for physical survival from predators but for survival and well-being as a humanity.

By living in our minds—the past or the future—we forget to live at all. For life is right in front of us, happening all around us.[56] I finally realized I was not living in the present because I was afraid of life. Anything to get away from the present moment because I was terrified of the world and its brutality, so I didn't want to be here. It can be really tough out there! Especially for sensitive souls. Many master teachers from the past have said that they would not choose to come back to the Earth/human experience at this time. It's a tough one but we all *chose* to be here. And we are tougher than we think. The more we remember that we chose to come to experience this reality, and that the "good" and "bad" are simply aspects of learning, the fear dissipates. Instead invite in love. For I am well. I am safe. I am at peace.

DIRECT YOUR MIND OR IT WILL BE DIRECTED FOR YOU

Our thoughts (and therefore our perception of reality) are shaped by what we allow into our minds, what we focus our attention on, and what we choose to direct our minds to. It is time we took back control of our minds and brought it to our presence—our essence, our solar plexus… our breath, our life.

This is as simple as starting to be aware of what you put your attention on and, when you catch yourself focusing on something that is negative or not part of the world you want to create, switch your focus to what it is you do want to create. I have a roll of images I keep in my mind as backup

---

[56] Of course we must consider the future when we make decisions for that healthy continuity. But I believe that when we truly begin to *live*, we will be equipped to make smart and harmonious decisions with mindful, heart-centered intelligence and discernment, for the greater good of all beings.

in case I enter into an onslaught of media[57] and need to quickly reset my mind. Keep your stash of images simple so you can pull on them anytime. Mine is a big juicy yellow lemon. Imagine for a moment, the bright yellow color, feel the rough texture of its skin, smell the zesty aroma and the bitter kick when you bite into it. Can you taste it? It focuses my mind again, ignites all my senses, and brings me back to my center. It also trains my mind to be able to hold a focus.

We as a collective can no longer focus our minds. Ram Das called us "drunken monkeys". If we are going to manifest an amazing world and work our magic, we need to be able to hold that vision for more than two seconds! Train your mind to be strong, focused and directed, to hold a vision of what you want to create. My good friend Manari often reminds me to focus on what I am creating, without letting a single bit of doubt or negative thought in. Just hold that focus, stay open to your connection and believe. You are extremely powerful.

I was once asked to present at a conference with Indigenous communities and government officials in Patagonia, Argentina with one of our board members—a brilliant, tough and incredibly inspiring woman from the Tahltan Nation. We had flown fourteen hours and arrived in this small-town hotel where we were booked to stay. It was one of those old hotels that was built when things were still done slowly, by hand with care. It was sturdy and had very thick wooden doors with big bolts on the doors. I woke up on the morning of the conference and took a shower. As I got dry from the shower, I went to open the door and the door handle fell out… on the outside of the bathroom door. The door latch was well inside the exterior wall, so this door was not going anywhere. I assessed my situation. This

---

[57] Media today—whether it be the news, TV shows, movies or ads—are largely filled with lower frequency content and perpetuation of fear. I won't pay to have my frequency lowered, on purpose! Let's demand better.

was a tiny washroom with no windows that was already steaming up, and the main door to my hotel room was quite far away, with a dead bolt.

So, there was no chance between all these solid wooden doors that someone would hear me if I yelled out for help. I sat on the floor and started to think for a moment. If they started to wonder about me at the conference, which didn't start for another hour, someone may come to check up on me, but the hotel would call first and not get an answer, probably assume I went for breakfast and was running late. If they did finally decide to check in my room, they would have to break down the door as that bolt was big! So, by my calculations, I was stuck there for at least three hours but likely closer to six or seven, with very little air and no phone. I would also miss this very important conference and likely be passed out, naked on the floor, by the time someone found me.

I started to panic, for about ten seconds. And then stopped, took a deep breath, centered myself, and then felt this immense sense of calm come over me. I realized that if I panicked, I wasn't going to get out. I needed to work with Great Spirit—connect to the energy of the door and this room, and work together with these energies to find a way out. I calmed myself down, opened to my connection, and said—with full faith, belief and confidence—"Great Spirit, I am getting out of here in the next forty-five minutes, and I will make my conference in time, so please show me the way".

ALL OF A SUDDEN, IT WAS AS IF OBJECTS IN THE BATHROOM BECAME ILLUMINATED

First it was my make-up kit with my bobby-pins. I bent one open and tried to screw part of the handle back on with my bobby pin. It didn't work. Next, I was directed to the metal cup holder attached to the sink so I tore it off and used it as a latch under the door to pry it open. I got half an inch. Next the towel rack which I tore off the wall and made it into a crowbar. Some movement, but not enough. Then came the toilet roll holder and one thing after another in the washroom. Slowly, I was able to pry the door open with one object while I pushed down on the latch in the wall with another. It was getting there but still nothing. Time was running out. This was a very important conference and we were the main speakers.

I paused, re-centered myself with three deep breaths and with full confidence said to Source, "Okay, here we go, I am getting out now". I pulled the door one more time with my fingertips and it readily flung open. Just like that, as if it was waiting for me to just get the combination of energy right. I thanked Source and ran out, quickly got dressed and made it to the conference just in time. I let the hotel manager know politely on the way out that they may need to send a crew up to um, do a bit of cleaning up. And that they definitely needed a new door!

We had a very engaged discussion at the conference that day about the rigorous environmental, social, economic and legal policies and frameworks Indigenous communities in Canada had created and that were needed before any development was entertained in and around Indigenous lands. It was a big hit and we made some great relationships. I would have missed it entirely if I had let fear in and stopped believing 100%, even for a moment.

I have found that many times it is not the first thing that works or even the second, third, fourth or fifth. It's about finding just the right combination, the key that unlocks the safe. You have to keep trying until you hear the 'click'. It ultimately comes down to belief but because we are all still novices at using our minds to create intentionally together with the energy around us, I find I have to sometimes build up momentum in order to strengthen my own belief and resolve (and to let the Universe know just how serious I am!). Then, the simplest, smallest action will be the one to push the energy into physical manifestation.

I was once at the airport around the holidays waiting for a flight to Peru on a project we were inaugurating in the high Andes. The Minister, the Mayor, the President of the communities and my business partner Jerry would all be attending. My flight schedule had a few stop overs so it was critical that I make this one out of my hometown of Toronto. Of course, Air Canada overbooked the flight as I learned is typical practice to account for no shows, but because it was the holidays, everyone showed and I was bumped to the waiting list. And because it was the holidays, all the flights to Peru were booked for days. I would never make it in time.

I went to the counter and politely explained my situation to the flight attendant and asked if there was anything that could be done. He looked at the list and apologized but said there was really no chance of my getting on since I was number 38 on the waiting list and everyone had shown up. He could see I was very disappointed, so he offered that if I could get someone to give up their seat for the next flight out in a few days, with the $500 they were offering, he would let me take their seat. Okay, I had a plan!

So, I started asking everyone in the lobby if they would be willing to make the switch. It was a room full of Peruvians on their way home for the holidays so, there were no takers. I was getting desperate. I actually stood up on a chair and, in my most moving Spanish, delivered a speech about the work we were doing with the Indigenous and building more sustainable economies in their country, and that I was expected there to launch a new and important program with the 77,000 Indigenous Peoples living in the high Andes. I did get some applause, congratulations and thanks, but only apologetic expressions. Still no takers. But I could feel the energy building, so I kept going a while longer.

When I started to feel tired and a bit defeated, I decided I needed to regroup. I went to the coffee shop next to the lobby and sat down for a few minutes taking a number of deep breaths. I was able to regain my center and my sense of calm, which brought back my faith and belief. I stood up with new resolve and commitment, ready to make it happen. As I stepped out of the coffee shop into the lounge, the flight attendant I had spoken with over an hour ago walked up to me and gestured behind his back. He discreetly handed me a boarding pass, smiled and walked away. Just like that!

    He had been watching me "work the deal" and decided it was worth bumping me to the front of the waitlist. Someone ended up cancelling and one seat opened up. He gave it to me. I would not have gotten it had I not built up the energy to say to the Universe, "I am serious about this! I won't give up! I will keep on going!" It was such a great lesson for me because it really showed me that you have to be committed, *work the deal*, keep at it... and miracles will happen. Whether it is small things like getting a seat on a plane or bit things like persevering with building an organization, a movement or a heart-centered business, holding the vision and believing in its success has been a secret sauce in my life. Even when it seems that the world is conspiring against you, that there is no hope, that it is the end of the line, that is when you have to double down on your beliefs, dig your heels in and go for it!

ANOTHER TIME, I HAD BEEN IN THE AMAZON

We had canoed all day to get back to the main village where a car was waiting for us to take me to the bus stop to catch a bus to town, after which I would have to catch yet another bus to the capital for my return flight to Toronto. It had been a long work trip and very rewarding, but I was ready to get back home, and there were only three flights a week leaving from the Quito international airport to Toronto.

    As we were getting close to the village in this rickety canoe, my friend—an Indigenous leader from the community—says to me that the river is dried up ahead, so we would have to dock the boat and walk the rest of the way. I was in thick black rubber boots to my knees, shorts and a t-shirt, soaked from trekking all day through the rainforest. We loaded up our things and began walking. By the time we got to the other village, our ride was long gone. We had been almost four hours late and the sun was starting to set so the driver must have assumed we weren't coming. There was no cell phone reception, hence no way of calling another car to come, and there were no cars or any motorized vehicles in the village.

My Indigenous friend looked at me and said, "We walk". I asked him how long it would take to get to town, and he said about two hours. Not only was I exhausted from hiking and canoeing for the past 8 hours, soaking wet, with a backpack full of stuff, but the last bus was departing town in twenty minutes! I would surely miss my flight.

WE NEEDED A CAR AND WE NEEDED ONE ASAP

So, I put my bags down and I said to my friend, "We are getting a ride". He looked at me somewhat confused but then understood what I was doing, so he put his bag down next to mine and waited to see what I would do next.

I looked up to the Universe and said, "We need a ride to town please, and we need it within five minutes." I then sat down and waited. Three minutes later, some fishermen who had taken a wrong turn came tumbling down the street in their truck and stopped to ask if we needed a ride to town. They said they came fishing there often and had no idea how they got lost this time! I just smiled and winked at the Universe. My friend laughed for he knew—the Universe works in mysterious ways. We got to town just as my bus was shutting its doors. I hopped on, thanked my friend, gave him a hug goodbye, and made it to my flight just in time. Trust and faith are a hard thing when you are so close to the wire but those are the times when it is most important. It is when the "miracles" really happen.

These are relatively small things. Imagine some grand things we could create by all focusing our minds *together* on what we wanted to happen. Quan Yin talks about the power of energy that is created in the mass consciousness during Valentine's Day when all minds focus on the perception of love. She says that it creates a current of high-frequency energy that blankets the Earth. Even though many people don't really understand that word "love", it is a step towards bringing it into the emotional body and the mass consciousness of humanity. What if we all focused on creating clean oceans and vibrant forests? Abundance of healthy food? Cars that run on air? What else could we co-create together?

INVITATION

Start to feel it. What does it look like? Sound like? Feel like? What colors are there in your image? What smells? Tastes? Use your imagination. Describe it to a friend and embody it. Become unapologetic about your vision and commitment to building a better world!

If we don't know yet what it looks like exactly, it's okay; we have the greatest teacher of all—nature! We can look to nature for inspiration. This is called biomimicry[58]—highly efficient and decentralized systems of water filtration, cooperation, creativity and renewal. In order for a forest ecosystem to work, a tree brings up water, shares sunlight, stores carbon, purifies water and maintains a multitude of shelters for wildlife.

This planet is an engineering marvel. And we are a part of that. We are not separate from it.

---

58 Brief intro to the concept and opportunity of biomimicry—learning and designing/building inspired from nature by my friend Jamie Miller at Biomimicry Frontiers. (https://youtu.be/Nx_wDAlsLP0)

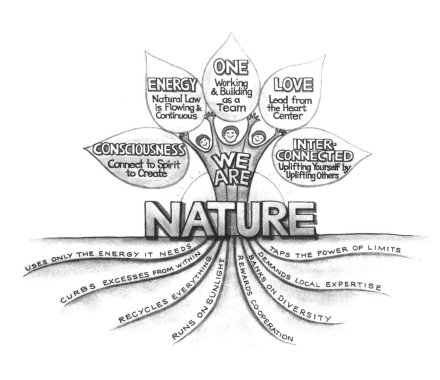

WE ARE NATURE

So, we are all interconnected in this big web of energy, making us one with all of creation. That means we *are* nature, not separate from it. As nature, we are also hard-wired with the same knowledge and intelligence required to maintain healthy and balanced life systems. But we overrode the system (or at least we are working really hard, stressing ourselves out every day, to override it! Why exactly?). We tune out, we ignore our inner voice, we resist the connection. This is how shamans in the Amazon I have worked with are able to condition the environment as they like. They know that there is no separation. They connect and engage with the energies around them. They open their minds to the knowing that there is no separation.

I was visiting some friends in the Amazon a few years ago. A very powerful shaman in the Shuar Nation invited us to join them for a week of sharing knowledge, ceremony, humor and adventure. It was spectacular. My boyfriend and I slept on wooden planks in a hut that had no walls but was dry at night which is all we needed. All week, we experienced the Amazon as I had never experienced it before. Butterflies, birds singing, no mosquitos or cockroaches, no pumas or jaguars coming into the village. It was very serene and peaceful.

On our last night there, the shaman left to go into town to prepare for our visit the next morning. That night, we were awoken by a gunshot from one of the villagers to ward off jaguars coming into the village. We then noticed that there were mosquitoes and cockroaches all around our hut and even a few snakes slithered in. I asked the shaman the next day what had happened to create that change in experience. He just smiled and said, "I like my home a certain way". His home is the forest. He conditioned the space, and when he left, it went back to being the wild Amazon.

We can all do it. We have just not trained our minds or strengthened our belief in our inherent *nature*. We depend so much on outside sources for solutions, for answers, rather than standing in our own power that's been inside of us all along. Many of us already condition our favorite room in the house to feel a certain way by bringing our energy into it—decorating it, making it feel comfortable and cozy, just how we like it. We also know when a place we visit just doesn't feel good and we choose to avoid it. We can expand this understanding and apply it to all things in a conscious co-creator way. What would the world we want to create feel like?

On a recent birthday of mine, I had found a recipe for these incredible zucchini chocolate muffins—vegan, gluten and sugar free, and they were delicious. I wanted to serve them at my birthday gathering I was hosting with friends and family. But my birthday is in the middle of September, so the zucchini plants had stopped blooming in the garden. They had bloomed so well all summer, providing us with such abundance for which I was very grateful, and had built a great relationship with the plants throughout the summer, spending time nurturing and caring for them.

So, I went out to the garden, sat down and had a talk with the zucchini plant. I said, "I have very much enjoyed the beautiful harvest you have provided for us this summer. You know I am very grateful. I am having my family and friends over this weekend and I really want to share your bounty with them. Can you please provide me with just one more zucchini? The next morning, I went out to the garden and there was a giant zucchini—the size of my arm laying there in the middle of the garden! It was the biggest one we had harvested all summer. I smiled and thanked the zucchini plant (a.k.a. Great Spirit) for this gift and I made enough zucchini chocolate muffins for everyone to have some to take home with them. When you interact and give in gratitude, you receive in abundance.

WE CAN COMMUNICATE WITH NATURE

Nature tries to communicate with us all the time. We *chose* to forget the language of nature and the principles of natural law that are built into us as a part of the ONE I AM.[59] We were all once able to speak the language of the trees, the birds and the frogs. We have stopped listening. The language is found when we pause and connect to our heart-center. It is within reach for all of humanity because we are nature! Duh!

All it requires is for us to listen.

[60]

---

59 See footnote 34 for an explanation of "ONE I AM" if you missed it!

60 Taimane, Mother (Earth)

I take every opportunity I can to practice—you know what they say about learning a new language—practice makes perfect! So, I ask the big hanging maple leaf on my morning walk to tell me about how it lives and what it has seen. I listen and feel for the response. Often the pine trees ask to have some of their needles taken for tea as it is very healing for my eyes they say. Always so thoughtful and loving. But then they all want to share some of their medicine so I take a bit from each tree. And I say thank you. In the mornings, I say 'good morning' to the sun, sky, to the wind, to the clouds, to the trees, to the plants, to the birds, to the grass and to Mother Earth. I wish them a good morning and thank them for being there. It is my small way of grounding myself in the knowledge that we are all part of nature and they are all a part of me and me a part of them. Acknowledging this warms my heart and makes me smile.

ELDER WHABAGOON BLESSING:

Count the number of Flowers Blooming in Spring and speak to Root Nation and Plant Nation. Kneel gently before them and become their friends. Take the time to build a relationship with the plants and listen for their song. Whisper sweet nothings and share your laughter with the plants, and they will share your intent with all their relatives that live down the lane. Notice the community they have built around them. Be kind and gentle to these new friends.

Nature is always communicating and we are excluding ourselves from the fun. We research and study plant and animal behavior when we could engage and simply ask how it works. What is fire? What is water? What are all these things that we were part of creating and which are a part of us? And if nature is so incredible, *we* must be incredible! Look down at your hands. What a marvelous creation!

One time, I was walking with my drum around a clearing in the forest, beating away and chanting, radiating loving energy. Out of nowhere, a mother deer with her baby ran toward me with this expression as if I was a long-lost friend that she was so excited to see again. It was the most beautiful thing! She got all the way to about a foot away from me with her baby in tow, when all of a sudden, I let fear in. I allowed my mind to think, "Isn't this strange... with her baby? That's not normal. Usually they protect their babies. What is she doing? I don't understand it; therefore, I am afraid of it." And just like that, I knocked myself out of my higher frequency of love and acceptance, to fear and separation.

At that instant, the mother deer looked at me again as if for the first time, screeched deafiningly, and bolted for the forest with a look of pure terror for her life and that of her baby. It was as if I "tricked" her into thinking I was one of them—nature in true form—and then took my mask off to reveal that I was still "human mind". It was an incredible lesson. Nature perceives by feeling. It can sense us, and we are the ones that are separating ourselves from that world, that beauty, those relationships.

A few months later, I was meditating on my yoga mat on the cliff next to my cottage overlooking the lake. I must have been out in my meditative state for a while. All of a sudden, I hear shuffling next to me. I open my eyes and sitting right in front of me on my yoga mat, is a big beautiful fox, curiously staring at me. I had learned my lesson with the deer so this time I just sat there calmly, maintaining my state of being. I said, "Good morning!" to the fox, and it looked at me with its big eyes. I was aware somewhere inside of me that foxes can be quite vicious, but I was not afraid. We sat like this for a few minutes, conversing without speaking words, and then the fox got up and casually walked away back into the forest. Five minutes later it was back with its buddy and the two ran around playing in front of me as if we were all old friends. It was beautiful and simple at the same time. Comfortable and natural. Just like being at home.

Now this is not to say that nature can't be cruel and out for the kill. Yes, everything needs to eat and that is the game of life and death. But within that there is also incredible kindness, compassion, generosity and cooperation. A grizzly may kill the moose and be the first to get his meal. But then the eagle benefits, then the foxes and the rodents, and then the microbes in the soil. The nitrogen-giving plants give and then get back. "A moose may give his life for us. When we die, our bodies may go into the soil (or so it was originally intended) and a tree may grow. In death there is life. Everything is a circle".[61]

We are part of that circle of life. We are part of the land and we will go back to the land. Our fear is preventing us from having the most incredible experiences that were the whole point of this existence on this living library called Earth.

LIFE IS EXPERIENTIAL—LIVE IN WONDER AT THE SHEER MARVEL OF CREATION

We are missing the awe, the brilliance, the vastness that is the beauty of the natural world. I was sitting in my living room with the new wild barn cat that I had adopted, for the first time venturing into my house as he spends most days and nights on the prowl outside. We had a mouse problem in our home in the country, and I didn't want to lay traps or poison, so I decided to get a barn cat. Some cats are wild due to being strays or abandoned. This cat grew up in the wild, living outdoors, and it was all he knew. So, he much prefers playing outside.

But like any Canadian autumn, the cold was setting in and the nights were frigid. So, I looked at Levko ("little lion" in Ukrainian) and asked him why he doesn't prefer the warmth and comfort of the house on such a chilly, wet night. He just gave me this look that said, "Are you kidding? The outdoors is so much more vibrant, exciting, multi-dimensional and fun. Inside is boring. I've already seen all the rooms and there's not much going on in them!" I laughed and knew he was right.

The next day we were fixing the water trough in the horse pen and left the gate unlocked for a split second. The horses have never attempted to leave (and always have the option of jumping the not-too-high fences if they really wanted to). All of a sudden, they bolted through the unlocked fence and ran! They rolled around in the grass,

---

61 Jerry, when I asked him about the natural balance of life.

they kicked their heels and they had the time of their lives! I chased them until I was completely out of breath, worried they'd run into the road or into the forest. Finally, the leader of the group, Catilla, stopped and, looking at me, said, "We are just having some fun and enjoying the taste of freedom for a few minutes. We won't go far, and we'll go back when we are ready". I said "Okay" and I let them play. Forty-five minutes later, they stopped, turned around, and calmly walked into the pen.

I KNEW IT WAS THE UNIVERSE SHOWING US THE BEAUTY OF TRULY BEING FREE AND WILD

All animals (including humans!) want to be able to experience life. In that moment, we decided to create a space at the farm for a horse sanctuary so that rescued horses could roam freely and experience that beautiful taste of true freedom. Just like Levko. How many of us have truly experienced absolute *freedom*? Without concern what people think or say? Without worrying about what we look like, how we dress or act?

A few years ago, I was driving up to a community in the mountains of Peru with a friend, a young leader from a small and isolated community high up in the Andes. We started to ascend a steep part of the mountain, where it goes to 5000 meters above sea level very quickly and I started to get light headed, as was par for the course each time I did this trip. I would have trouble breathing until I acclimatized to the lack of oxygen. I was curious if my friend who grew up in these mountains could feel the sudden difference in air pressure. I turned to him and asked if he could feel any change as we ascended. He looked at me with a big smile on his face and said, "Yes, I can breathe again". It was the most incredible statement that has stayed with me to this day, for we were now far enough from the city and back in nature, that he could truly breathe—referring to the state of peace and calm, connection and *freedom* that he felt when back in nature.

> ### ELDER WHABAGOON BLESSING:
>
> *Oh, our beautiful Mother Earth! Step gently on our Mother Earth. Respect each step, for each step is a lesson and we must honor each step we take upon her. Our steps are the lessons, and they are hidden under every leaf and rock. We are to live in harmony with Mother Earth. She has taught and shown us the ways through the natural laws and cycles. Look upon Mother Earth as being very sacred and kind.*
>
> *The Tree Nation that bring us fruits and nuts, the plants bring the beans and squashes, and our waters provide us with fish and water. She generously provides for us, so why is it that we can't look after her better? As young people, you will now be going to guide us through the next generation. It will be up to you to take care of Mother Earth, to heal her, to help her flourish. Our way of life is living in harmony with the winged-ones, the swimmers, the fliers, the crawlers, the Root nation, the Plant nation, and the Tree nation and our Grandfather Stones, the four-legged and the two-legged, for we are all one, and we must come back together as one, to walk the Sacred way.*

#SAVEHUMANBEINGS

Our relatives, the Elder trees, the root Nations, the bee societies, the Grandfather mountains and the governing oceans, are doing their job. They continue to clean the air we breathe, pollinate plants for food we eat, provide the freshwater we drink, and regulate our climate, as was designed. We humans have forgotten our role within nature's beautiful intelligence. When we remember who we are, we remember we are one. When we destroy what is outside of ourselves, we know we are destroying ourselves. What we see around us in our world is a reflection of what we are doing to ourselves. We are the devastation and destruction we are creating. *Earthquakes, tornadoes, droughts, wildfires—heart attacks, cancers, stress and anxiety*. The world is an incredible teacher because it teaches by mirroring back to us our actions and behaviors. And it's waiting for us to wake up and pay attention to the lesson. It keeps hitting us over the head. *Earthquakes, tornadoes, draughts, wildfires*. "Anything? Nope they're still asleep... must have been a big night out last night to sleep through that one!"

We do not need to "save" nature. This is human arrogance. We must respect nature and build a new relationship that reflects the respect, gratitude and compassion that we ourselves desire. Mother Earth can heal herself if we give her a rest. In three days, she would breathe in and out, and rejuvenate herself; the oceans can restore themselves; the soils replenish; the air purify, *if we allow them some space to breathe*.[62]

In fact, during the COVID-19 pandemic, where planes stopped flying and cars were not on the roads, the bugs came back, and the skies were blue again. Nature (including ourselves) has had an opportunity to go inside and clean

---

62 The same goes for our own bodies BTW. Our bodies are incredible engineering marvels that are designed to heal themselves.

house. It does not take much: just slowing down enough to breathe and to let Her breathe. It is part of the healing journey we are all collectively on right now. We must remember how to *be* again. Life can be brought back to wholeness when we bring ourselves back to wholeness, for we can only truly save ourselves. When we begin to create from a place of consciousness and awareness, the world around us will reflect that thought—that New Story we create.

WE ARE WORKING AGAINST NATURAL LAW

We work so hard for what we think is "abundance". But we have it backwards. It's not about working hard, being stressed and exhausted. Ever watch a hawk glide on a windy day? It is one of the most majestic things to watch. They don't try to go against the wind, rather they flow with the wind and allow it to take them where it will go. They are open to the ride and enjoy every moment of it. Imagine them trying to control the wind and telling it that it wants to go the opposite way? That's why we are burning out. That's why we are so stressed. That's why we have to work so hard. Because it is not natural.

Nature *flows*. A river doesn't have to work hard to go downstream; it just does. We are trying to flow upstream against all powers and natural forces. And it is literally killing us. We have cut ourselves off from our life source and wonder why our (human) machinery is starting to have glitches in the system. It is time we let go and have faith in ourselves. Because when we trust ourselves, we put faith in the Universe and send the signal loud and clear that we are ready to co-create. We still need to steer the ship, but in the flow of natural law.

WORKING *WITH* NATURE

When we create according to natural law,[63] as governed by the heart-center, we have limitless possibilities. Which means healthy competition and cooperation. Ever wonder why wild meat tastes so much better than stuff we get at the store, even if it is organic? The wild moose meat I eat up in Tahltan Nation territory in northern British Columbia literally makes my body feel alive: a surge of energy courses through me—there is nothing like it. There is an Indigenous tribe in Kenya who believes there is a natural exchange that occurs when we hunt in the natural way; that through healthy competition, the animal will choose to give up its life to the fair victor, and in so doing, pass on its life-force energy, as opposed to animals kept in captivity—without the choice, without the natural chase. We are the ones who lose out.

Yes, nature can be vicious, but if you observe closely, it knows the rules of the game. A bear will not run the salmon extinct. The wolf will not kill all the deer and destroy their habitat.[64] When we start to work with nature, it will work with us. You get what you give. And right now, we are not giving much to the rest of our nature family, and the energy we are giving is that of greed and destruction. So, it's what we are getting back, IN ABUNDANCE.

In a report developed by a circle of Indigenous Elders,[65] its authors present the Indigenous view to redefining "conservation" as relating and engaging, learning

---

63 Which means just make it work for everyone. At the very least, don't destroy the whole. Be an individual, prosper and thrive! But don't take down the community in the process. Seems pretty fair, right?

64 Because they seem to like to eat good, healthy food. Every year. Interesting concept.

65 *We Rise Together*, The Indigenous Circle of Experts' Report and Recommendations, March 2018, p.35.

and growing together, rather than as excluding and limiting activities:

> *Indigenous worldviews differ fundamentally from the philosophies that guide many Crown-protected areas, where conservation is achieved by restricting activities and limiting access. In Indigenous worldviews, conservation is achieved when the relationships and uses that have conserved the lands and waters for thousands of years remain intact or are re-established.*

They go on to say,

> *Indigenous languages and place names, as well as knowledge systems and laws passed down through the generations, provide oral records of Indigenous Peoples' relationships with their lands and waters. The health of the land and of the people cannot be separated. They are interdependent.*

In her book *Braiding Sweetgrass*,[66] Robin Wall Kimmerer, a member of the Potawatomi Nation and world-renowned botanist, sets out the example when talking about the harvest of sweetgrass. Through scientific study and investigation, Kimmerer discovered that the same regions in the United States where sweetgrass was going extinct were the same areas where the Indigenous People had stopped their traditional harvesting of sweetgrass for basket weaving. Conversely, in areas where the harvesting was continuing for traditional uses, the sweetgrass was thriving. Through various scientific simulations, she was able to show that plants and, in this particular case, sweetgrass, *want to be interacted with*. They want to be

---

[66] Kimmerer, Robin Wall. (2013). *Braiding Sweetgrass*. Milkweed Editions.

harvested; they want the relationships and engagement. Just like any being of energy, when left alone and ignored, sweetgrass begins to wilt and die.

• INVITATION •

We need to take down the fences of our natural parks that we have put up in our minds and get out there. Engage, interact, learn and stomp your feet on the Earth! The land wants life on it. It's crying out for it. Go and get your hands dirty. Let the Earth know you love her and that you are grateful for the abundance she provides. Let the water know how thankful you are for its subsistence. Thank the trees for the air you breathe every day. Thank the bees for working so hard, especially in August when we are all on vacation: they are the busiest of all! Look around, observe, acknowledge, interact!

STANDING WITH MOTHER EARTH

I personally love this planet and will not support an agenda to go live on Mars if we destroy our life support systems here on Earth nor will I support depopulation agendas that say there isn't enough for everyone. Earth is our home and it's our world to create. There is abundance for all if we work in harmony with life systems. The Amazon supports trillions of life forms and there is no welfare or unemployment. We created a mess and we can create beauty. Giving up on Earth is frankly a cop-out and insulting to Mother Earth who has taken care of us and continues to do so.

Let's not be lazy. Let's take responsibility for ourselves and humanity, and let's all agree that it is pretty awesome, the potential we have on this beautiful Earth. Let us demand creativity and connection, love and understanding—from our education system, from our government, from corporations, from our families and friends.

Most of all, this means that *we* need to start looking at *ourselves*. How are *we* thinking? What decisions are we making? How do *we* treat ourselves, each other, and the Earth? Our thoughts create our reality over time. If we think unhealthy thoughts, how can we expect to create healthy realities around us? It cannot be. We have created an ugly world because of ugly thoughts that we allowed to create our reality.

Let's begin to treat ourselves with love, compassion and kindness; let's treat each other and the planet with the same. It must come from the inside out. Let's take back our stage. It's our play and our world. Let's collectively set some good intentions and put some good thoughts in our mind funnels. Building towards beauty. Towards light. Towards a world that lasts on planet Earth and that works for all. Let's get to work!

THE FIRST STEP IS SELF-RELIANCE

This is not financial. This is being able to rely on yourself. In times of uncertainty, chaos and crisis, it is more important than ever to build and foster self-reliance. Food, energy, water. Greenhouses can be built for less than $200. A garden can fit in anyone's yard. Foraging is available to us all. Simple, clean, decentralized energy systems are getting better and better every day. Water filtration systems are now so compact you can take them camping!

We are so dependent *on* "the system" that it is hard to have the confidence to stand up and create something new. We have allowed ourselves to depend on the system for our basic necessities, how can we really say "no" to it? The power is in your hands. Support local farmers, grow your own food, start to forage in nature. It is all possible, even in cities. That is true power. Because it is true freedom.

TRUE HARMONY IS POSSIBLE

I had the opportunity to sit in meditation with my friend Manari on one of his visits to Vancouver from the Amazon. He led me through a meditation where I went down into one of the remaining intact ecosystems that exist in the mountains of British Columbia. The animals all came to greet me, and I felt the soils, the richness of the minerals, the purity of the waters, the vastness of the mountains and the intelligence in the plants and wildlife. I will never forget this big, white mountain goat that came up to me and began to play with me, as a dear friend. He was so funny, we ran and played together. Somehow, we were able to communicate, and he showed me the balance that existed in this life system.

It was so powerful and so beautiful that when I opened my eyes again, I had tears streaming down my face, but with the biggest smile, from the sheer beauty that I had just experienced. Manari looked at me and asked if I took the opportunity to go deep into the core of the Earth, if I soared across the sky and saw the mountains from above, if I felt the clouds. I looked at him and said, "No, I was having too much fun playing with the mountain goat!" We both laughed. But that feeling of total harmony with life, that intimate relationship that exists, that we are a part of, has stayed with me since that day.

INSTRUCTIONAL MANUAL FOR HOW TO BUILD A RELATIONSHIP WITH NATURE[67]

(1) Notice it.

(2) A few days later, smile, wink or let it know in some way that you are interested.

(3) Dig down for the courage to say "Hello".

(4) Ask it if it wants to… umm… maybe… hang out sometime.

(5) Plan to spend an afternoon together. Go for a walk and make sure to ask questions (so as not to seem too self-absorbed—first impressions are important!)

(6) If things are going well, maybe you "accidentally" brush up against each other.

(7) When the time is right say, "Thank you for a lovely time", and "I hope to see you again soon".

(8) After a few dates and when you are ready… take a deep breath… and say, "I love you".

Nature wants to build a relationship with you. It is waiting for you to make your move. It will embrace you, and it will be a friend that you can count on for life.

---

67 For beginners, best applied to dating a tree or a bush. A coyote or other wily animal is not recommended in level one as they like to do their own thing and often play hard-to-get. It can also be applied to dating yourself, at an advanced level (because we are the toughest one of all! Not very nice to ourselves!).

CREATING HEART-CENTERED REALITIES

We are born of love and will all go back to love. We spend our whole lives looking for love—to be loved, to feel loved, to give love. Yet we fear going deep into our hearts and allowing ourselves to fully feel and express the pure unconditional love that lives there. We fear standing in our love—our greatest strength and power. It is the greatest force in the Universe, and it is our greatest weapon as Heart-Centered Warriors. Darkness, fear and hate have no chance against love.

LOVE

Love is our true essence. It is us without the fear, the guilt, the shame, the pain. It is our purest selves and it is when we feel most like ourselves that we see glimpses into our true soul.

> *"Love is the total absence of fear. Its natural state is of extension and expansion, rather*

than comparison and measurement... When we help ourselves and each other let go of fear, we begin to experience a personal transformation. We start to see beyond our old reality as defined by the physical senses and we enter a state of clarity in which we discover that all minds are joined, that we share a common Self, and that inner peace and Love are in fact all that are real. With Love as our only reality, health and wholeness can be viewed as inner peace, and healing can be seen as letting go of fear". [68]

---

[68] Gerald Jampolsky's explanation of Love in his book, *Love is Letting go of Fear*.

Love is not something to 'get' outside of ourselves. We all have it and it's right inside of us all the time. In fact, we all possess limitless amounts of it within ourselves. It has simply been buried beneath a whole lot of mess for a long, long time, so we forgot it was there. Let us take off the layers—the guilt, the shame, the fear, remove the covers from our hearts and let them be free to shine.

• INVITATION •

When you feel fear, shame, guilt or any of those uncomfortable feelings, close your eyes, breathe into it, feel it, settle into it, and begin to fill it with light, let go, breathe. What we resist persists, when we sit with it, and we fill it with love, it will release and allow the love to shine through once more. Try it now.

THE FUTURE IS FOUNDED ON A FOUNDATION OF LOVE, NOT FEAR. PERIOD.

It is the only way that we will continue on this planet. It is abundance, not scarcity. Kindness, not hate. Wisdom, not hypocrisy. Leadership, not narcissism. It is a place where everyone is standing in their own true power, living their true essence, contributing their unique value to the whole of humanity. To do our part, we must remember who we truly are. We are powerful energy beings, co-creators of many Universes. A part of nature, not separate from it. We have been duped into believing we are lesser than this. But, in fact, we are greater. We are pure light and love, even if we don't feel that way right now. It is a journey of taking off the layers to reveal our true selves. Underneath the self-doubt, low self-worth, shame, guilt and fear that have been placed upon us—underneath it all—we shine bright.

In this time of fear around the world, when you raise your frequency and choose a higher path, whether it is by smiling at someone rather than snapping at them or giving yourself a nice healthy meal over fast food, you will begin to see that energies of fear have no real power. They are inherently weak. Their only power is that which we give to them. In the face of love, hate and fear will shrivel up and die. While it may be scary to open your heart and radiate love in the face of darkness and fear, like any path less taken and life's greatest journeys, it is for the brave and courageous, and it is what makes superheroes of us all. Now is the time—for the Earth and for Humanity to shine bright!

OPEN YOUR HEART AND KEEP IT OPEN NO MATTER WHAT—IT IS THE SOURCE OF YOUR SUPERPOWER!

Love is our true superpower and the natural fuel of the Universe. It is the strongest and greatest power source because it is the energy of Creator. It is what gives life and it is necessary for life. Babies without love in orphanages will often get sick or die within months. It can bring us to life, and it can bring all things to life around us. It is our own free energy source, available to each one of us, *in abundance*. With this potent energy, we can create abundance wherever we intend and share that powerful energy with all beings around us. There is no shortage of abundance on this planet.

Nature in the Spring is a beautiful thing to watch. The sun heats up the water in the plants and in the soils, and it is that heat that changes the molecules of the plants to begin to come to life. We as human beings, just like the plants, are made mostly of water (approximately 60%).

When we open our hearts, we are warming our own bodies that change the water molecules and we, too, begin to come to life. And with that life-force activated within us, we can create thriving life around us.

A few years ago, I had just read Dr. Emoto's book on tuning water with heart-centered messages and music, and I wanted to test out the theory. Since all life is made of water, I decided to try it with a bouquet of 12 roses I had just received from my boyfriend standing on my counter facing the window. They were beautifully arranged on my counter facing the window with the sun beaming through. I decided each day to tell one rose (the same one daily) that I loved it and pay it special attention. The rest I would simply ignore. Within days, the roses began to wilt and soon all died. Except the one that I had given attention to. It lasted in full bloom for over a month. I have since tried this experiment in many different ways and it never fails. Give it a try yourself!

• INVITATION •

Stand tall, breathe the light of the Universe into your heart until it is overflowing with light. Then send it out to the Earth, into the land, the plants, the trees and the soils; to your friends and strangers on the street. Send it across the globe to all nations and people. That is the energy that brings life. And we can only harness that power, that free power, by going to our heart-centers.

We've all put triple-quadruple bolts over our hearts, and so our life-force energy is suffocating—we are suffering, as is our world. The energy is trapped and cannot flow. We have done this to ourselves and can undo it ourselves. We don't need gurus or religion to save us from this suffocation.[69] We can simply intend to do so and breathe love and light into and out of our hearts, allowing them to release and open, until it feels safe to come out. Fill them up with light like a bulb getting brighter and brighter and brighter until you are blinded by the light. This is our gateway to the Everywhere Spirit. No middleman needed. We all have a direct line.

As an ambitious lawyer, I had hardened myself and had bolts around my heart — all business, all the time. Then, a few years ago, I attended the *Speaker Adventure* weekend workshop founded by my friend Jeff Salz in California, a true adventurist. I wanted to improve my speaking ability and be more comfortable on stage engaging with people, which I was not at all at the time. We had to prepare a speech ahead of time that we would practice with throughout the weekend. At the end of the weekend, we would deliver the speech on-stage, in front of an audience, recorded as a full performance.

I showed up Saturday morning to this beautiful space set out in nature, and the speaking coaches proceeded to lead us through an entire morning of heart-opening exercises. It was not what I had been expecting! I was ready to simply delve into my talk—time is money! They said the exercises would help free us up to speak with ease.

So, I went with it, curious to see where it would lead. We did breathing and meditation exercises, saging and a beautiful ceremony to open our hearts. By the time noon rolled around, we were all feeling amazing, like we

---

[69] Not that there is anything wrong with gurus. I have had A LOT of spiritual advisors over the years—all of whom I love and admire greatly. Gurmukh, often referred to as the Queen of Kundalini Yoga, once said at a kundalini gathering in Toronto, that the age of gurus is over—we must now be our own gurus. Seek the answers within ourselves. I thought this was quite a beautiful and humble statement.

were floating. We had literally spent all morning bathing in our open hearts. We began the speaking part of the workshop and I was called up. I think to myself, "I got this! I've practiced, I have a good speech prepared and now my heart is open, so, great—no problem!" I get up in front of the group of eight other participants and our coaches, and I dive in.

I get about four sentences in when one of the coaches jumps up, yells, "Stop! Stop!" He runs up and grabs my speech right out of my hands. I was frozen. "Did I say something wrong?", I ask. He looks at me and says, "That was your *mind* talking. Where is that beautiful heart we just opened? I don't want to hear a dry recitation of facts, I want stories, color, vibrancy, experiences! I want to *feel* your passion! Now tell us all a story of one of your adventures. You have been around the world, worked with incredible people and traversed remote jungles. That's what we want to hear about!" I was stunned but deep down I knew that he was right. I had retreated back to where I was comfortable, where it felt *easy*.

This open-heart thing was new territory and I was uncomfortable playing in it. But he wasn't going to let me get away with it. So, I thought about it for a moment and then began to share a story, then another, and another. The coach had me share four stories in total and then another three over dinner that evening with the group! That night, as I lay in bed, my whole body was buzzing. I felt more alive than I ever had. I felt the warmth coursing through my veins. I was so energized from the life-force flowing out of my heart, that I must have only slept an hour or two, but I felt more refreshed and full of energy the next day than I ever had. When I had to get on stage in front of a full audience and a professional recording team to deliver my new speech the next afternoon—a series of stories of my adventures and work around the world which I had written that morning and had no time to rehearse,

I nailed it! No notes, just heart. And it flowed and flowed. It was the first time I experienced the power that exists within our own hearts.

Now, this may take some time as our hearts have been in hiding for ages. But be patient and keep at it. You owe it to humanity and most of all to yourself. As the systems around us begin to support life instead of suppress life, we will feel freer and freer to be ourselves. When we release our hearts from this solitary confinement to which we have relegated them to, all our beautiful heart energy will lift the frequency of ourselves, those around us and the whole planet. We will automatically begin to rebuild our relationship with all life, including the Earth, the animals, one another and ourselves. Animals can sense us from very far away. With open hearts, they sense their relatives. With closed hearts, they sense our separation from the Earth family.

This is not a 'lovey dovey' thing as some of my friends would say. The most Heart-Centered Warriors are the strongest and toughest there are. Mother Theresa healed with love and kindness. But she went into the toughest places and was strong as nails. Indigenous women who have carried their communities through the atrocities that persist today, carry themselves with grace and resilience as they nurture and spread love and caring to those in need. It is a soft energy, not a hard one.

We have had thousands of years of war, hate, fear, anger and aggression in our human story. The way forward is through love, kindness, generosity, compassion and lots of hugs! We all have a lot of trauma to heal. Lifetimes of it. We must all now stand in this world with grace. Grace is strength and gentleness at the same time. Strength is unwavering knowing of who you are, while acting from the heart-center with love and compassion and kindness for humanity, knowing we have all suffered tremendously.

WHAT IS THE "HEART-CENTER" ANYWAY?

The heart-center is definitely an elusive term, much like "love". Often used, rarely understood. It is the place within yourself where you feel—in complete peace and harmony—with yourself and the world around you. This place cannot be reached through the thinking mind. It is the mind of your true self. And it hides in boredom. Not rushing, anxious, trying to prove something, do something, accomplish something or win. When you are confident and trust in yourself and the Universe, you let go and allow.[70]

In our heart-centers, we are pure love and pure light. There is no fear, no jealousy, no anger, no hatred. It cannot exist. It is true beauty; it is divinity. The communities I had the blessing to work with high in the Andes of Peru, had been through some of the most traumatic experiences—from terror attacks to drug trafficking rings and violence, to big mining coming in and changing their lives overnight. Yet, they continue to wear their hearts on their sleeves. They continue to show the vulnerability of being who they are. Yes, there are many challenges they face. But beneath it all, they shine their beauty and are proud of who they are. Their honesty and sincerity is inspirational to all who they meet.

We have all felt at one time or another that we could not, by society's norms and standards, stand in that vulnerability, yet that vulnerability is our strongest power. And it is time now to take off the layers and stand in that power. It is the power that can move mountains, unite people and allow us to do impossible feats.

CREATING ABUNDANCE FROM HEART-CENTER (A.K.A. ENERGY BLOCKCHAIN)

Giving from the heart-center means giving for the sake of giving, what it is you have within you to give, and the more you give, the more you receive. This is another one of those natural law principles. When we give our love

---

[70] This is not laziness or not being ambitious. In fact, some of the greatest inventions in the history of humanity were created from this state. Einstein did his greatest work after a walk in nature meditating. Creatives Mozart and Monet both meditated and allowed genius to *flow through them*.

unconditionally to others without the expectation to receive, the love within us simultaneously expands. It is the law of great abundance.

There is no difference in the energy world between material objects and human actions. So, by giving love, without expectation, through a smile or a hug, through financial support or a gift of a home cooked meal, it is all accounted for as *love*. It is 100% fool-proof accounting—all on the open ledger (energy) network. There are 7 billion users. We can all relax and trust that the system is keeping track of every transaction. And imagine the abundance we could create by all taking part in that network! So, love away!

It gets even better. Energy always seeks the void (because it's always seeking to reharmonize to wholeness). That means if you give what you have to give, without expecting anything in return (giving in its purest sense, which is the strongest form of energy), you will be brought back abundance of what it is you need, when it is you need it. So, if you have a need for money but you have an abundance of kindness and compassion, if you give that kindness freely, it will find the person who needs it most. The money (which is just energy) that is in excess will find its way back to you when you need it most. This is sacred economics.[71]

I was on a blind date once, set up by one of those online dating apps I tried out.[72] It was right around the time I was reading Charles Eisenstein's *Sacred Economics* and quite taken with his concepts. While the date was a bit of a bust, my date shared the most incredible story that brought tears to both of our eyes. I knew then that the Universe connected us for me to truly understand the sheer beauty of this sacred energy system. He shared the story of his trip as a student to Syria with some friends years ago.

---

71   Charles Eisenstein, *Sacred Economics*.

72   I mean, who hasn't been just a little bit curious about those apps!

They had gotten lost in a small town and had trouble finding their way back. They had blown through their cash and had no way of getting back to Damascus. They started hitch-hiking, and a young man from the town stopped and generously brought them to his home, where his family fed them and gave them a place to sleep until the bus the next day. The next day, the young man proceeded to drive them to the bus station, paid for their tickets, and gave them some money for food. The students made their way back to Damascus and went back home to tell the story of their adventures.

Nearly two decades later, Syria was faced with a brutal war and refugees fled from the country. The group of friends, no longer students but successful professionals, decided to sponsor a family of refugees to come to Canada. When they went to meet the Syrian family at the airport, they greeted them and the family thanked them, and then inquired what led to their generous gift. The friends replied that they were helped selflessly when they were in Syria years ago and in desperate need.

As they proceeded to recount the story to the family, the father's eyes began to tear up. He said that years ago, he had helped some stranded tourists[73] and housed them for three days, fed them and helped them on their way, without asking or expecting anything in return. At the time, his family had a successful business in their town and could afford to help some strangers in need. He went on to say that his father told him after the guests had left, that he had done well, that he was proud of his son, and that one day, when he needed it most, there would be a helping hand. His father had since passed away and he had forgotten about his father's words until that moment. Now his family had nothing and was forced to flee their country. And help found them when they needed it the most.

---

73 No, not the same group—that would be a story out of Hollywood! But what hospitality the Syrian people have!

THE AGE OF SERVICE

Wayne Dyer says in *The Shift* [74] that trusting in yourself is trusting the wisdom within that created you and is all-knowing. He goes on to say, "Forget about yourself and reach out and serve". That is the pathway to true happiness. Lao Tsu said that the four main virtues for any person are: (1) reverence for all life (respect); (2) sincerity (honesty); (3) gentleness (kindness); and (4) service to others. We are entering the age of service. That is one the highest states of being and yet the simplest. It is not "charity". Service is much more practical than that. The more you give, the more you receive.

My dear friend Mary, an Elder with a heart of gold, works tirelessly in service to others, helping youth at risk and homeless in the city. She never asks for anything in return and lights up the room when she walks in. Her smile is infectious and her laugh is contagious. She is one of the most beautiful people I know. It is the vehicle through which you create infinite abundance in your life and for all those around you, thus creating an abundant and healthy planet.

• INVITATION •

You may start with simple actions—slowing down to open a door for a stranger; making eye contact and smiling at another; sending love to someone sad.

True leadership is service. Jerry told me a story once about their Nanok, the highest leader in their traditional governance system. The Nanok was responsible

---

[74] *The Shift*. Directed by Michael Goorjian, Lyceum Films, 2009.

for ensuring the well-being of all the people and of the community. He would eat last, only after all had eaten. It was not a paid position. It was an honor to be able to serve with generosity of spirit. The more you gave, the more you were respected. It was a recognition of the value of intangible wealth. Intangible wealth are the values that all people espoused to be able to pass on to the next generation. It included the values of how to be in this world, how to live well with the natural world, how to create value for others, stand in and seek the truth, and live with integrity, honesty and humility. If one only received tangible wealth from their Elder, it was considered less valuable than the intangible wealth that made a person truly great. And if the Nanok was no longer serving for the well-being of the people, he would lose the privilege to be the Nanok.

We have forgotten this in our society and no longer hold our leadership to account. As a result, our leaders no longer strive to be great leaders—to build good character and lead from values and a service mindset. As Simon Sinek stated in his Ted Talk (2014), "The job of a leader is not to be in charge, but in taking care of those in our charge."

All these actions inspire our hearts to wake up and get into the driver's seat once again. There is no lack of human suffering on the planet. The greatest lack we have today in the world is that of love. Children are born and raised without love and Elders die alone. In traditional Ghanaian culture, no one is without a mother and father. If a child loses a mother, another is given to that child. In this way, everyone has a purpose, and everyone is loved and cared for.

SELF-LOVE

Yes, we are going to talk about this! It is the most important thing we can all learn to do right now (and it was my biggest struggle!). How to do it you may ask? I know I did. I literally had no clue. I started by giving myself kindness and compassion. Be easy on yourself (no, not unambitious or apathetic... but *nice*). Stop beating yourself up. Jesus said, "love thy neighbor *as yourself*". Self-love is at the top of the ladder. It is the hardest and highest lesson because you cannot serve others until you can fill up your own tank with love. "Attention, attention: put your own oxygen mask on first before assisting others". It is not selfishness. It is the opposite. It is the life-force within you that allows you to blossom so that you may give love freely to others.

It starts with self-care and self-nurturing. It is quite literally the "food" that nourishes your body. "The Buddha said that nothing survives without food, including love. If you don't know how to nourish and feed your love, it will die. If we know how to feed our love every day, it will stay for a long time. One way we nourish our love is by being conscious of what we consume. Many of us think of our daily nourishment only in terms of what we eat. But in fact, there are four kinds of food that we consume every day. They are: edible food (what we put in our mouths to nourish our bodies), sensory food (what we smell, hear, taste, feel and touch), volition (the motivation and intention that fuels us), and consciousness (this includes our individual consciousness, the collective consciousness, and our environment)".[75]

Without self-love, one cannot grow into maturity. It is for each of us our own inner magnificence. In the presence of that magical energy, no darkness can exist. It is the ultimate lesson because it is the full and complete acceptance of who we truly are. No other person can heal another. They can only facilitate the journey back to self-love. To attach our love to material things, we are not learning to love ourselves and others. Self-love gives us the deep knowledge and understanding that everyone deserves love, and every being is worthy of it.[76] It is the glue that holds together our ultimate harmony with self and all beings.

---

75 *How to Love*, by Thich Nhat Hanh, one of the best-known Zen teachers in the world today.

76 Quan Yin as taught by Shin Yin.

Thich Nhat Hanh reminds us that:

> *"Love is a living, breathing thing. There is no need to force it to grow in a particular direction. If we start by being easy and gentle with ourselves, we will find it is just there inside of us, solid and healing....Each of us can learn the art of nourishing happiness and love. Everything needs food to live, even love. If we don't know how to nourish our love, it withers. When we feed and support our own happiness, we are nourishing our ability to love. That's why to love means to learn the art of nourishing our happiness"*.[77]

So, give yourself a nice body massage, place your hands on the soles of your feet and feel the life-force energy going into the soles of your feet, thanking them for supporting you all day. Sooth your eyes from the strain all day by rubbing your hands together to generate energy and place them over your eyes, your head, your heart. Feel the pure energy and light moving through you. Walk with bare feet on the Earth. Sit and read a book under a big old tree. Nurture your mind, your body and your soul. Feel your true self coming to life. You are beautiful!

Bliss (I Am the Light of My Soul)[78]

---

77 Thich Nhat Hanh, How to Love.

78 Kaur, Sirgun & Singh, Sat Darshan. "Bliss (I Am the Light of My Soul)". *The Music Within*. 2011.

THE JOB OF HUMAN BEING

"The essence of being spiritual is learning how to be consciously human".[79] Watching the workers at the hotel I was staying at in Mexico was beautiful. They do their jobs, always with a smile and such pride in all they do, whether their job is sweeping floors, serving cranky tourists or leading exercise classes. They put their hearts into it and radiated such joy and positive energy that was contagious with all those around them. They made me realize that no task is menial if you do it with all your heart and put your whole self into it. Our jobs are one vehicle through which we may shine our light to the world. It is our opportunity to share our love and light with the world. And that is a beautiful thing. Because that energy infuses everything and multiplies. These are true energy masters! Radiating love and light to even the rudest tourists. For $5 a day. So, I tip them well. Not because they performed their job technically well, but because they performed the job of *human-being* extraordinarily well.

HAVE THE COURAGE TO GIVE THE GIFT OF LOVE

Yes, it is very scary to open our hearts at first because not everyone will reciprocate right away. Every soul is at different stages of being able to give and receive love. And that's okay. It's about giving the gift of love without expecting anything in return. Love without expectation. It is the most freeing way to love.

Clan Mother Pimastan, a beautiful soul, will sit at coffee shops and beam out love to heal the room and the people in it. When I am running late in a meeting, she says to me, "Not to worry, I will go over to the women's shelter with my drum and spread some love to uplift the hearts there while I wait for you to be done". When I go to pick her up at Union Station in downtown Toronto, she is always

---

79 Matt Kahn.

standing out front confidently with her big drum, beating away, singing, with a huge smile on her face, spreading the love to all the people that pass her. She stands firmly in her light and spreads it every chance she has, even if no one reciprocates. That makes her one of the most courageous people I know.

We've got the most powerful force in the Universe inside of us. How might we all start to be okay to feel it, share it, embody it? What if we feel hurt because we shut ourselves out, afraid of the powerful light and beauty that is inside each one of us? Our love is not only our life-force, it is also what is going to heal us, our communities and our planet.

• INVITATION •

Look people in the eyes and send them love. Sit at the coffee shop and radiate love and light to all that walk in. Say I love you to water, to trees, to the worms! Because we are all one, loving out brings loving in.

LESSONS FROM THE DOLPHINS

We as a human race are novices at love. We fear it and resist it. Yet it is our greatest force and the core of natural law. We can learn a lot from the dolphin species, who represent the one I AM through the lens of freedom in the expression of pure love. They show us what freedom, joy and companionship comes from by allowing ourselves to love and be loved fully. In fact, scientists have studied dolphins and discovered that their brain cortex that is responsible for the feelings we associate with love are three times bigger than in humans! They have really *developed* that capacity. What capacity are *we* developing? Like any muscle, what we work at we get better at and grow our capacity and strength in. Are we ready to take on the development of this highest and best of all abilities? The one ability that will make us stronger and more powerful than all the Universes combined?

THE BIG LIE

We have created a fear-based society that stalls energy and prevents thriving. No wonder we all feel stressed and sick a lot of the time! The big lie is that if we were to create well-being for all and a thriving planet, then it would mean we would all have to do with a lot less individually. This is only true in a society based on the zero-sum game, which we have all allowed ourselves to believe is the only way we can live. But this is not natural law rules! It is man-made and totally against the natural way, which is built for scarcity and limited potential.

When we live from a heart-centered reality, we create for abundance and infinite potential. We create abundance and the economy is 10x, 100x, 1000x bigger! Because we are all living from our true essence and this is limitless creativity, innovation and ingenuity. Love is the frequency of the multiverse, where all the wisdom, knowledge, energy and potential reside. When we connect and tune in to that frequency, it's like tapping into the motherboard of motherboards. Anything becomes possible and potential is limitless. *We* are limitless.

LOVE = ABUNDANCE. FEAR = SCARCITY

The energy of *love* is the highest vibrational frequency there is and therefore equates with abundance. Fear, as its opposite, equates with scarcity. Love is the highest form of intelligence. It is harmony. It is flow. When you create a system on fear, there is never enough; you are always needing more. Some have said that you can never have enough of what you don't need. Hoarding and holding on to what you have rather than sharing and letting the energy flow creates stagnant energy. In our bodies, this creates dis-*ease*. In society, it causes a break-down on a grand scale.

We have built our world and societies on fear and scarcity thinking—fear of breaking the law, of not succeeding, of losing our money. Fear that there will not be enough, so we must take more for ourselves; that in order to be rich, others must be poor. We have forgotten that we are all energy and everything is energy, that we are limitless, infinite co-creators. The world is infinitely abundant if we work with it through the lens of life-affirming energies. Because at the heart of natural law is the power of love.

Our human existence on Earth was never meant to be a struggle. Food grows in the forests and in the fields to do the work for us so we can focus on learning, growing, developing ourselves and our human experiences. Water was provided for us so we can nourish our bodies and thrive.

I always marvel when I spend time with some of the Indigenous communities who still live in the remaining healthy rainforests of the world. When we go out for excursions on projects, they never pack food or water for the day. Instead, they snack from the land throughout the day. Pick a coconut here, a few seeds there, a fruit off a branch or some sugar cane. Catch a fish, even some lemon ants for protein![80] Spending time learning, playing with the energies and *living*, rather than having food drive your day. Always in the flow, and not being afraid it will run out, because they know the rules of the game—when you nurture, care, respect and give, nature is abundant. The abundance we see in these rainforests is a creation of those minds. Those minds that believe in abundance and thriving ecosystems are holding that story for humanity. It is therefore no surprise that 80% of the last remaining healthy biodiversity on the planet is on Indigenous lands.

---

80 Yes, I did try some lemon ants, and they do have a lemon zesty flavor to them!

THE GREATEST GIFT

I was on the Mexican coast in autumn for a personal retreat (*a.k.a.* some time alone doing my own thing, writing, meditating, practicing yoga and tai chi, and just enjoying being with myself), staying at a cute boutique eco-hotel. I was feeling recharged and headed back to the airport to go home, and my flight got cancelled. So, the airline put us all up in this monstrous resort on the ocean, with capacity for 10,000 people. Nature was literally paved over for artificially controlled "beauty". Nature, but not too much nature, so as not to make people uncomfortable. There were a lot of issues with seaweed in the water at the time. My eco-hotel stopped spraying the plants and did their best to support the re-harmonization of the ecosystem, recognizing it was a sign that there was clearly an imbalance.

This big shiny hotel that I was put in for the night was the complete opposite. They literally put up walls to block out the ocean and beach pollution and built a water amusement park within the hotel. That night, I crawled over the wall and went to sit on the beach. I was the only one there. And suddenly, I had a very sad feeling come over me, that it was just that easy for people to forget about nature. It reminded me of the Dr. Seuss movie, *Lorax*, where we destroy nature and say, "Oh well, the ocean is too polluted to swim; let's go play in the water theme park instead". Just like that, we forget about that connection, and it is lost again, for perhaps hundreds or thousands of years.

In that moment, I had such a deep sense of gratitude for those Indigenous and non-Indigenous peoples from around the world that have held open that door, that memory, for humanity, of our connection with our natural world, so that even if we forgot for a brief moment, it would not be completely lost. That is pure unconditional love for humanity. I was in awe of the magnitude of that gift.

CREATING A HEART-CENTERED ECONOMY

I challenge us all to change to a system of love which provides abundance for all. Anything and everything is possible in this new reality. We can easily sustain all the souls on this Earth, if done from a place of life-affirming principles and adhering to natural law systems. Yes, it would require us to think and consider a great deal more. A forest sustains infinite life forms because it is highly efficient and works within natural law. It knows the rules (for the sake of healthy continuity) and plays within them (so as to be healthy and continue). What a crazy idea!

LOVE BASED ECONOMY TRAITS:	FEAR BASED ECONOMY TRAITS:
Abundance	Scarcity
Leads to more	Leads to less
Connected	Disconnected
Accretive	Zero-Sum Game
Inclusive of All	Exclusive of Most
Everyone busy in healthy ways, contributing	Unhealthy busy
Entrepreneurs fulfilling real needs	Entrepreneurs creating more things we don't need
Creative	Mechanical
Natural	Unnatural
Life affirming	Life destroying
Balanced	Unbalanced
Fulfilled	Unfulfilled
Healthy, living in joy	Unhealthy, living in stress

Which reality do you want to foster?

My good friend Lawrence, the pioneer organic dairy producer in Canada,[81] developed a very successful organic dairy company that truly fosters these love-based traits. For almost 40 years, he has been working to ensure that not only the welfare of people is taken care of through environmentally friendly, healthy organic products—from beginning to end, but that the animals are also cared for and loved. Each of their 200 cows has a name and is truly part of the family. Everything from the climate, to the pasture, to the type of materials they walk on, is done in consideration of animal welfare. The cows even have massagers to keep them relaxed when they have to be indoors during very cold spells in the middle of Canadian winters! As a result, his cows produce milk for years longer than conventional dairy cows. When they do finally stop producing milk, they go to the "retirement village" at the farm to live out their lives in peace—to wander the fields and sleep in cozy bedding, rather than to a slaughter house. "They earned it!" as Lawrence likes to say.

I have always been so impressed with the thought and consideration that goes into their work. Yes, it is a great deal more to consider, and hard work to make sure it is done right. But the tangible and intangible rewards are infinite. It is truly a heart-centered business. Some time ago, some scientists tried to measure the difference between Lawrence's dairy and conventional dairy in a lab and concluded that the "ingredients" are the same. But what they missed was the intangible—the healing power of the love your grandmother puts in her soup; the value of animals raised in a loving environment; the value of those relationships founded in respect and reciprocity. Science has yet to be able to measure these intangible aspects of wealth, but they are what lead to a truly healthy and happy life.

I love visiting my friend Becky Big Canoe at her home in the summer. She is a true Indigenous matriarch, and she

---

81   Harmony Organic.

fosters community values well. As we sit having a cup of tea and chatting on her front porch, other women drop by for a conversation, to bring over some honey or jams, to talk about their families or challenges they are having. Women from across the country find their way to Becky's house on a small island First Nations reserve because it is a place of comfort, of friendship, of compassion, of generosity, of love. And with all the challenges she herself faces each day—as a strong, tough Indigenous woman today—her door is always open, she always has some warm tea to offer, and she is always willing to lend a helping hand. This is the kind of neighbor I want in my community.

The ancient Indigenous way was that everything was done based on thinking and acting *today* in a way that will be sustained for seven generations. Many harvesters take only 50% of their crops, leaving the rest for the animals, for the soil to replenish and the seeds to grow for the next harvest. We at our home garden have a 60/40 rule with nature—in their favor, and they respect these rules ...most of the time! We are the only ones that have the scarcity mentality, and so we take much more than we need, and we hoard, cheat and steal. But we end up cheating ourselves, because as the birds are playing and coasting on high winds, feeling the joy of creation, we are stuck in an office or stressing with our work—never enough, always more. It is the mentality with which we have approached life with until now. So that is what life gives back to us, in abundance. The good news is that we can now choose to no longer experience suffering and instead choose joy, choose happiness, choose fun!

When we connect to our wholeness and live from that state, the world around us will reflect that as well, because it reacts to the energy we put out. That is why the Indigenous knew the secret—that drumming and singing beautiful love songs to water would leave a most beautiful imprint on the water. What energetic imprint are *you* leaving on the world around you as you interact with it?

> ### ELDER WHABAGOON BLESSING:
>
> *Embrace your water. Love it, respect it, Say miigwetch to it all the time. Sing to it in the shower, sing to it while you are doing the dishes and while you are brushing your teeth. Talk to water when you hydrate the plants and speak to them about their connection with one another and tell them from a two-legged perspective what a beautiful relationship they have. Speak to the water when you place it in the various parts of your home. Water is never far from you. It is beside you when you go to bed and its the first thing you reach for as you sit up. Water is never far from your feet. Water is life, water is our life. We are all protectors of this Sacred Nibi. Sing to it, respect it, embrace it, love it. Miigwetch Nibi.*

Ne-be Gee Zah-gay-e-goo
Gee Me-gwetch-wayn ne-me goo
Gee Zah Wayn ne-me-goo

Water, we love you.
We thank you.
We respect you.

SLOW IT DOWN TO SPEED IT UP

Many Elders and shamans say that our sickness in the West has been caused by going too fast. We have quite literally forgotten how to stop and smell the roses. For the past 100 years and especially the past 20 years, we have been like a race car going at 4,000 miles an hour, revving our engines, and we have literally burnt ourselves out. Steam is coming out of our sides. The problem is that once we are on the high-speed track, it is so hard to slow down or to stop as we get used to the next rush. I know: I was a major adrenaline junky, after the big deals in corporate law and running on a couple hours of sleep, coffee and... adrenaline.

I hate to tell ya, but our bodies can no longer sustain the stress we have put on ourselves, without our true power-cord. I learned that the hard way. And since the outside is a reflection of what is on the inside, our natural environment is reflecting back to us what we are doing to ourselves. In a state of wholeness, illness and sickness cannot exist because we are *whole*. Building a better world must be created from the inside out—from a place of wholeness.

When we remember that we are all part of the one and all interconnected, we remember to slow it down in order to speed it up. Rain falls one drop at a time, yet overtime can shape mountains. When we work together and flow as one with all of creation, we can create greater abundance for all. When we resist natural law, we have to work so much harder to get anything done. And the results are, well, what we have today—a lot of people stressed, exhausted and with little hope for the future. So, breathe out a big sigh of relief. It is all good, we've got this and everything is OKAY.

In order to create harmony and balance in our world, we must garner guidance and inspiration from our higher selves, from the Universal Mind. This requires us to slow down to hear the guidance. Many of the greatest inventions, discoveries and breakthroughs in our history came from this place. Nikola Tesla said, "My brain is only a receiver. In the Universe there is a core from which we obtain knowledge, strength and inspiration".

When we create from a place of connection, we have the greatest library of knowledge available to us. When we create with our 3D mind alone, we are limited to the ideas, notions and knowledge that we have been exposed to in our lifetime, or perhaps that we are able to read in the library online or in our city. Needless to say, this is and will always be a more limited approach when compared to the Universal Mind, which includes knowledge and information of all that has been, is and will be, for all time.[82] Not only can we garner solutions for our many challenges; those solutions may include aspects and components that had not previously existed in the material realm.

WHAT IF WE COULD THINK A CAR INTO EXISTENCE?

Everything in our physical world is a material manifestation of what first exists in the energy world. We are, after all, energy beings having a human experience. Think about it. When you have a great idea about something you want to create, you first think of it (mental body). If it passes the test of being a good idea, we get excited at the prospect of creating it (emotional body). Then we begin to visualize it becoming a reality (spiritual). Then finally, we create it into form (physical). The idea is already developed by the time it hits the physical plane!

What if resources did not need to come out of the ground? What if we strengthened our mental capacity and ability to simply think a car into existence? This requires a complete unlearning of what we believe we know, and a re-learning of what we have forgotten. We do not need to exploit and destroy to "advance" as a civilization. In fact, I would argue that this is not advancement. *Advancement (in my opinion) is to harness the true potential of humanity to be co-creators of a sustainable world that works for*

---

82   Yep, that is one heck of a library!

everyone and every being. That is a true challenge; that is true complex thinking. We must work the deal to get that one right!

Human potential is endless and vast. It requires us to tap in, to go into our heart-center and feel. And to get there we need to *slow down*. But it is a fallacy to think that slow means less advanced. If slowing down means being able to discover relativity or paint the water lilies or invent an airplane, or grow a strong tall tree trunk like a big Oak, then isn't it the secret to the greatest achievements that humanity has ever known and the most intelligent designs of nature itself? Perhaps we weren't ready before on a collective scale. But that time has come where people are waking up. Imagine the world we could create if we worked at it all together, including with our natural relatives?

It means engaging with life in a conscious way. Open yourself to all your senses and that is opening to *more*. Become an explorer again. When you walk down the street, notice the trees, the flowers and the birds. Notice the changes in the wind and the directions of the sun throughout the seasons. Pay attention to the people around you and the carvings in the wooden table at which you sit. Feel the food you eat, and think about where the ingredients came from, about the farmers who labored to grow the vegetables, about the chickens who gave their lives for you to have your lunch. Say "good morning" to the sun and take in that beautiful life-force energy. Talk to the squirrels in the park, the pigeons and the ants. What lessons might they have to share? Pay respect to the tree that has provided shade to your lawn for 60 years or 200 years. Imagine the knowledge and intelligence there. Let's begin to retrain our minds to open to a vaster experience and we will unlock the secret to true *manifestation*, *awe* and *joy*.

> **ELDER WHABAGOON BLESSING:**
>
> *So beautiful this Fall day. Look no further than the animals and nature outside your front door to see what you could be doing to embrace this beautiful season. Harvesting and gathering begins. Watch Squirrel with his cheeks full of nuts and running back and forth from his home. The leaves are beginning to prepare Mother Earth's blanket for the upcoming winter so please leave the leaves on the ground. The leaves are food for the Plant Nation. Leaves provide housing, nesting and bedding for wildlife and insects. Tree's will shed their leaves to continue a life cycle. It is only humans that come along and interrupt this beautiful cycle. These are all signals for us to purge, rotate bedding, check my winter boots and coats to ensure I am ready for the upcoming seasons. So, take a moment for yourselves and family to watch the wildlife and nature's habits to learn the lessons of preparing but also how to live together as one to walk a sacred path.*

Become a conscious observer of your world. Experience the world in a multi-dimensional way. Connect to the people around you, to the energy around you, to the ground and to the sky and everything in between. But most importantly, connect to *your self*—to your breath, to your life-force energy. Feel the breath. Feel your body. Sit and be still. Enjoy yourself.

Your source of connection is in your heart. Breathe into it. See it filling up with a bright white light. Then feel it extending out, radiating out. It is part of the greater whole. We are all interconnected in a beautiful white light of energy beings and consciousness.

• INVITATION •

When you feel alone, overwhelmed, stuck or suffocated by the reality around you, sit and breathe. Then breathe some more. Fill up with your bright white light. Wrap yourself in a big cozy blanket of this light. Then, try connecting with another life—an ant, a beetle, a tree, a bird. See a bit of yourself in the other, for this helps us to get a different mindset—we begin to truly feel the oneness with all of life. We begin to realize there is no separation, we can never be separate. Our support group is infinitely vast!

FEELING IS THE NEW SEEING (A.K.A. SPARKLY-DUST SUPERPOWERS)

The good news is that we are waking up by the millions and the young people out there today—you are wired for healthy continuity, which is awesome for all of us. What a relief! There is literally a sparkle in you, and this creates a sensitivity that many are struggling with. But this is actually a superpower! It may not feel that way now, but you are built for the future, for the new world that is being borne *right now*. The new world is a feeling-based knowledge because our heart-center speaks to us through intuition. We can all tune in and feel things around us and that is knowledge. Everything is frequency and vibration. It can be pretty scary and unsettling at first, and, especially for the supersensitive beings out there, it's really tough!

Being able to *feel* is so important because it allows for real-time feedback loop of our relationship (and impact) with and on all life. We can feel when our brother, sister or friend are sad, having a rough day or a bit sick. We know even before they tell us. We are able to go and cheer them up or bring them some warm soup. What if we could do this with all our nature relatives? Rather than waiting fifty years until we have chased away all polar bear life forms from our perception of reality, this superpower allows you to feel the imbalance immediately when you tune in, and feel what nature needs to rebalance, and then can correct our actions right away. They will tell us, if we listen. Now, that is cool!

We close ourselves off from this advanced system of communication because we are afraid of being vulnerable or sharing too much of ourselves. Yet, so much more can be said and shared this way. Plus, it is way more efficient! The simple look of a loved one across a crowded room speaks volumes. Begin to trust your feelings and allow them to communicate *for* you.

That is how the natural world communicates. That is how we can communicate. It is clairvoyance and clairsentience. We are sentient beings. We don't need words to let someone know we love them, that they mean something to us. We all know how good it feels when another is truly missing us or happy to see us, compared to when they say the words but the feeling behind it does not match.

The greatest part about this knowledge and communication system is that it is individualized and specialized just for you. It will give you information on which apple in the pile has the most nutritional value for your body; it will tell you which vacation package you will enjoy most—for you specifically. It will tell you which option on the menu will leave you feeling most nourished. It will tell you about others and allow you to serve better, to work better in collaboration, to develop better as a leader, and to know which projects to put your energy into. Pretty amazing! Imagine if you could know exactly the right route to take to get you home fastest without taking a chance with GPS; if you could know which jar of soup in the grocery store had just the right mix of nutrients and life-force for you; if you knew how someone was going to react to you *before* you said what was on your mind.

• INVITATION •

Try this at the grocery store. It is a great way to practice our connection and inuition. I will hold one apple at a time and feel it. Which one resonates the most with you? Everyone will have a different experience, because we are all unique beings. I do this with all my food, and in this way, nourish my body in the most optimal ways possible. Try it next time you are at the grocery store!

Ever wonder why you get uncomfortable in a room full of people or overwhelmed by feelings all of a sudden? You are picking up on all the tumultuous energies around you. There is a chaos of energies out there! And it can affect your health and well-being, absorbing all that negativity in the world. That is why so many are drawn to numb out our feelings. But don't worry: this sensitivity is going to make you function in the new reality that we are co-creating. We simply have to make sure to draw very clear boundaries of what energy we want in our space.

It is time to put a big circle of light around you to define your boundary of your brilliant energy sphere and affirm it as yours. Fences make good neighbors!

INVITATION

Close your eyes, set the intention, and then visualize a big sphere of energy around you which is your boundary. Make it go a few feet from your body to capture your aura as well. Make it any color you wish. I put pink roses around mine which absorb any dark energies that may come my way throughout the day. Use your creativity and imagination! Yours could be a spaceship or a beautiful ball of light. What does your light body boundary look like? Trace it around your whole body. Try it now!

Never allow the world to prevent you from giving love. Giving love will never hurt you; it only makes your own light shine stronger (which makes you feel better, healthier and more vibrant, by the way!). Don't stop feeling others; don't cut yourself off because that is the future form of communication, and it is happening *now*.

KNOWLEDGE TRANSMUTED BY FEELING IS THE LANGUAGE OF NATURE

Nature communicates through feeling, through sensing energies and intuition. In fact, every plant has its own aura color, its own sound and frequency.[83] See if you can hear it! We are nature. We just forgot the language. Similarly,

---

83 Arzu Mountain Spirit.

each of our chakras also has its own color, its own sound and its own frequency too! It is essential for our healthy survival that we remember it and re-learn how to use it. Let's hone it. Learn how to work it. Don't be afraid if you can feel things, people and animals. Use it to practice. Talk to the seagulls and the squirrels and the geese (mind their attitude—the geese have an attitude with everyone—don't worry, it's not you!). And even if we don't know what to do and it feels so overwhelming and like such a daunting task to get back to just feeling good, there are small and simple steps we can take that will keep us on the right path.

We are retraining our minds to remember what it feels like to feel good. Like working out a muscle that is no longer seemingly active, it takes a few weeks or even months of constant repetitions with the weights before you even start to see the changes. But once that muscle starts to build, it remembers! When I first started, I could not feel anything—I had succeeded in numbing myself as to not *feel* by ignoring myself sufficiently. I had to work on reconnecting with myself in order to feel again. Have faith and keep going even if it doesn't seem like there is light at the end of the tunnel. There *is* light at the end of the tunnel! And the tunnel is not as deep and long as we think it can be.

The saying goes, "Fake it till you make it!" When I was starting out, I would sit in front of a mirror, look into my eyes and say "I love you" to myself until I actually could feel something akin to a warm sensation when I said it. For a long time, I felt nothing, it was so foreign. It took months and months! And the more I sat with myself, the more I realized how much I had neglected myself, how negative my mind was, and how that was harming my body and all I did. So, for six months I challenged myself: for every negative thought I caught myself thinking or saying, I would stop and immediately think or say the opposite (positive). And, boy, did I have to stop A LOT. But this process started to retrain my mind to be way less annoying. And my body started to open up, my fear of life started to dissipate, and I started to create more positive life-affirming realities for myself and those I was working with. We are our worst enemy.

HOW DO YOU GET LOVE IN YOUR HEART? (EXERCISE)

Guess what, it's already there! We just have to tap into our own fountain of eternal life. So, close your eyes. Take a deep breath. Go into your heart and imagine one of those dimmer light switches that you begin to move slowly from dim light to brighter and brighter and brighter... until the bright white light has flooded your heart. Picture your cat or dog licking your face and feel that warmth for them. Think of a grandparent, a best friend, and hold that image.

Now focus on your heart and feel through your heart that love. Picture the white light expanding out all around you. With every inhale, fill your heart with even more light. With every exhale, radiate that light out of your heart. Imagine that light surrounding your whole body, then filling the whole room you are in... then the whole building, the city, the country, the planet. Imagine ripples of this bright powerful light going through the world, healing and uplifting all the souls. This is your most powerful weapon. It is what will heal you and what will help you in the darkest moments. Practice it now. Know what it feels like so that you can get back to it when you need it most. Hold love in your heart and radiate its bright light around you. Darkness doesn't stand a chance. It will obliterate the darkest darkness.

Now think from this place for as long and as much as you can. We all deserve kindness in our day, and most of all to ourselves. Living from the heart-center is incredible strength, self-assuredness and not compromising on values. It is ultimately about being good to yourself, knowing who you are and standing in your true power. This is about as tough and kick-ass as you can get! When we live from the heart-center, our body responds. And we live in a place of inner peace and outer joy. And that ripples to those around us. *It is being selfish for our own good and that of the planet!* This is being a true Heart-Centered Warrior.

I recently wrote a poem after a meditation out in nature. I was really intimidated by the notion of writing poetry. But I was told it was a great way to be begin to feel other life forms through our heart-centers, so I decided to just go for it! I sat with the fire late one night and was inspired by the way it was able to dance and express itself freely. It remained in its true essence and embodied it fully. These flames were having fun, playing with the wind, the elements. It was its interaction with life that was the source of its beauty. You don't need to be a poet or an artist to write something. Connect with a being out in nature and give it a try! What is it showing you? What is it teaching you?

TWO OLD PALS

Oh, the fire burns so bright,
Like a hot summer's day, rising well into the night.
It hugs the logs in a tight embrace,
Knowing that it belongs nowhere but this place.

Then golden flames stand so proud,
No need to scream or shout or even be loud.
Their confidence is clear, in the knowing what to do
and who to be,
Without need for competition or pretending you see.

As the wind blows stronger, it adds to the fun,
The fire will not retire or even come undone.
The two play together like old childhood pals,
Carefree and innocent, like a pair of young gals.

The sun begins to rise and the fun must come to an end,
The embers of the once strong logs are all that contend.
To tell the story of the adventures in the night,
That did not go to vain because they played with all their might.

WALKING IN GRACE—EMBODYING THE EAGLE AND THE CONDOR

We all hold within us both "masculine" and "feminine" energy. It is the balance of these energies that creates wholeness and the ability to navigate life with grace. It is with grace that we will walk in a heart-centered reality. Grace is a beautiful thing because it is the strength of a sword with the gentleness of a feather, at the same time. And it comes from being so sure of yourself and who you are that you can stand strong in yourself and what you believe in—but not from a place of anger, judgment or attack: rather from a place of gentleness and softness. Because there is no need to be angry or judge another when you are grounded in your true self. They have their views and you have yours. We only get upset because deep down we feel our view being threatened. But when you are strong in who you are, there is no threat. No one can take it away from you. It is yours and you no longer have to defend it. This is grace.

And it comes from integrating the feminine with the masculine, which we have all allowed to dominate for many centuries. We are very good at the anger, judgment and aggression game. It is time to soften and allow the nurturing, compassionate, gentle side of ourselves to come in. For the good of our own health and well-being (when we judge others, we are usually hardest on ourselves, and I know this first hand!), for the well-being of humanity and of our Mother Earth. No, it is not a sign of weakness. In fact, I challenge everyone to begin to approach situations that they would normally get defensive in, with softness and grace. And you tell me what the harder path is to take. It is easy to throw a temper tantrum or yell and scream. It is harder to pause, breathe and respond from a neutral mind, with kindness and generosity of spirit. That is a "black-belt" and a true Heart-Centered Warrior of the New Human Story.

These next ten years are critical to have simple tools to get back to your path and not get pulled away by negativity or darkness that is all around. Know who you are and where you are going. Own who you are. Own your mind! Stay focused and disciplined on values of decency and humanity, getting back to your true light and then staying there for as long as you can. Even if it's five seconds a day at first, it will grow and like a snowball rolling down a hill, build momentum, and then it is a giant snowman before you know it!

It is easy to jump on the bandwagon of what others are doing and eat bad food, drink too much, or use drugs often—doing what feels good in the short term. But it is like McDonalds: it tastes good in the short term but will make you sick in the long-term. It is the same with behavior and values. Good values—kindness, generosity, compassion—are harder than ignoring people, getting angry, hitting someone, and honking that horn... at first. Feels good in the short term, but ruins your day or week, and just makes you feel crappy overall.

I'm not going to lie to you: getting up early and meditating every day, exercising, eating a healthy veggie burger or salad, making eye contact with people and smiling, and *making an effort* to engage and be kind takes a lot of discipline... at first. Then you reach the plateau of the mountain when, all of a sudden, it feels great and makes you so happy that you start to feel repelled by anything that *doesn't* make you feel good and healthy and happy. Because that is our true nature. You start to notice when you ignore your higher self for a day, your body craves a good session of deep belly breathing. Pure love and light is our true nature. We just put a whole lot of junk on top so we can't feel it or see it most of the time. But start to take those layers off and you'll see. You will shine so bright—brighter than the brightest star in the sky. This is the future, so start now!

When things get tough, go back to YOU. 100% pure and natural, no additives. Remember you are a powerful energy master. Underneath it all, YOU SHINE BRIGHT. That means pure love and light. NO EXCEPTIONS!

This is not religion or new age guruisms. This is YOU! You are your own guru! Allow your own inner light to be your guide and illuminate your path. Repeat out loud this powerful mantra:

I AM the Universal Elements of Consciousness

> I AM... the nurturing mother/ father that cultivates the provisions of the Earth.
>
> I AM... the abundant empress/emperor that satisfies the thirst of each of my wishes.
>
> I AM... the worthy warrior that ignites the admired flame of reverence.
>
> I AM... the lover that compassionately breathes the breath of affection.
>
> I AM... the honest communicator that freely chants the sound of truth.
>
> I AM... the insightful intuitive that shines with the light of understanding.
>
> I AM... the heavenly guru that realizes the glory of God is within myself.[84]

---

[84] Jowett, Geoffrey (2014). *The Power of I Am: Aligning the Chakras of Consciousness.* Studio City, CA: Divine Arts.

THE CO-CREATOR'S TOOLBOX

These are simple, go-to tools that I use to keep my compass pointing back to my heart-center—my true self. It is important now more than ever that we can readily find our Center, that we do not get pulled into negativity or fear-based thinking. It is critical that we keep our frequency as high as possible, as often as possible. Especially when times are tough! Because as the old system comes undone, there may be confusion as the new system (the "green shoots") rises.

So, get armed with a toolbox of ways to remember what path you are on, to battle the energies out there in a peaceful manner. You can use mine or find your own—whatever works for you! These are some of my favorites. Feel free to take what you need! And always remember, when in doubt, don't know what to do, feeling lost and without direction, ask for help from your spirit guides and angels. Everyone has a great big team of helpers and they are there wanting so much to help you. But they don't want to interfere, so we must ask. Ask for help. Ask for help with big things, little things, all things. I ask for help in work, personal life, help with conversations, help feeling better, or even help finding the right way home! Ask for help; it is there for us all. We do not have to do this alone.

FROM F U TO BLESS U

When someone cuts you off on the road, BLESS U![85] When someone insults you, BLESS U! When someone takes your spot in line, BLESS U! It is so important because it gives us a PAUSE—a freeze frame moment before we react so that we may *respond* rather than react. And while we could just count to ten, it is very hard to remember to pause and count in the heat of the moment. Using a simple go-to response creates an easy habit that we can remember and, on top of that, with those words (as words are energy), we immediately are both giver and receiver of a higher frequency, thereby amplifying our own energy and the energy around us.[86]

---

85   Matt Kahn talks about blessing others when they do something to annoy or upset you. "Bless U" is a simple, easy to remember, go-to technique that I use, and it works magic! Keep reading to find out why!

86   Like hitting a triple-word score in Scrabble!

With those simple words, we are able to transmute the energy from a toxic, low-vibrational state, to a much higher state in an instant. We literally become transformers of energy. What this does is puts us back in our heart-center. It gives us a chance at a good day. It is the most selfish thing you can do! That split second could be the difference between a bad fight and a peaceful day. This is re-training your mind to stay high, naturally.

When we engage in someone's negativity and allow that energy in, we take on other people's crappy stuff. We literally dive-in and take a swim through their bad day. We take on their anger, their stress and their fight with a friend. Now *their* bad day becomes *our* bad day, and many times you never even met the person! So, why are you so willing to take on their "stuff" and carry it with you all day long? Say "Bless U" and give yourself the gift of a clean and vibrant energy field.

It is most important when it is the hardest. When you feel really low and the whole world is crashing down around you, that's when it is so hard to do. It starts with one "Bless U" at a time. Even if you are faking that smile, cringing while you say it, it still counts! That moment of truth when you are ready to tear someone apart (physically or emotionally) is when you pull out all the strength you can muster and bless whatever is around you. It counts the most. It's easy when life is roses. When you are upset with someone, when we spiral in anger, we do a lot of damage to ourselves. Plus, you pick up a whole whack of bad karma. We already have lifetimes of it (the history of humanity has not been all peace and love!), so let's not add more that we will have to deal with later. It is literally re-training our minds to a higher state of being, and to stay there, one "Bless U" at a time.

Practice when you are driving—when someone cuts you off, bless them and wish them a good day. You will immediately start to feel better and your day will be better for it (and you won't have to come back as a frog in the next life!). The cool thing is that since we are all connected, when you shift your energy and send the person who is clearly having a bad day, blessings (we have all been there), it uplifts us and they can feel it, too. It may just make their day better. And perhaps they pass it on. That is how we work to create a safe neighborhood, city, country, and world.

I know some people are like "Well, who cares about making their day better? They just cut me off!" But imagine a world where everyone was feeling good (a lot of the time)? I know, from where we are now, seems like an impossibility. But if we could, we would create a world from *that* place, and it would reflect a much more harmonious and positive experience for everyone, especially ourselves. In order to move in that direction, we must start to retrain our minds to behave from a place that we want to be in—to reflect the world we want to live in, today, in the present moment.

This does not mean that you have to accept bad behavior from others. If someone does wrong, it is okay to let them know that it is not right and how it made you feel. However, there is a difference between losing your cool and communicating consciously. The former will ruin your day/week/month; the latter will create growth in you and learning for the other person. It is the path of the Heart-Centered Warrior.

CONNECT TO CREATE

We are all always connected to the entire Universe, to Source Energy. It is the "motherboard". It is where all codes live for our DNA, for the world's DNA, for all life—animate and "inanimate"... and for all creation—past, present, future. And we are always connected to it. Einstein knew this; Tesla knew this; Mozart knew this. We forgot that we have access to this incredible library of knowledge and creation.

We can never be truly disconnected, although it may seem that way a lot of the time! We simply have to open ourselves up to more, to a greater possibility. It's like we have blinders on, so our view is very narrow and limited. We are blocking out 90% of our experience, painting with one color and one paintbrush. All we have to do is shift open those blinders and, little by little, open ourselves up to that brilliant web of existence to which we are all connected.

OPEN YOUR MIND TO YOUR EVER-EXISTING CONNECTION

The way to get there is to begin to tap into our other senses. When we walk in the forest, walk *with* the forest not *through* the forest, just like a good friend (don't you just hate when someone is hanging out with you but is not really there? It's the worst and we can *always* feel it). Everything is a relationship, including with our natural world.

When you walk with the forest, what does it smell like? Are there tastes in the air of rain just fallen? What sounds do you hear? Can you identify them? Are the birds chirping to find their young, or is it mating season? Acknowledge the community that exists there, between the trees and the animals and the birds, all communicating and working together to maintain life in balance. Ask for permission to enter their house. By doing so, we open up to the magic that is there. Look for the magic and you will find magic everywhere. All you have to do is look. When you walk in the park, don't just *look* at nature: *feel* the wind, *hear* the birds, *smell* the grasses, *taste* the fresh air. We are multi-sensory beings and the "motherboard" is a multi-sensory mainframe. To open to more, we can begin by *actually opening* to more! More senses, more experiences—more, more, more!

Ever wonder why life feels so hard sometimes? Well, think about it. We were given this incredible, limitless energy source: the ability to heal, create, experience, live in joy and wonder. But we cut ourselves off. We are running on fumes rather than on our innate energy. We have cut ourselves off from the life that we are a part of. Plug back in and you will get the true 5D experience!

Right now, we are watching the movie in black and white and wondering why we aren't getting the full effect. Experience the movie with all your senses. 5D means moving *beyond* only what we can see and touch. It is also what we can feel and what we can create. Interact with it. Add to it. Play. Engage. *Be* the nature that you were *born* to be!

CREATING FROM JOY

The place that you create from is what sets the intention and energy for your creation. Do you build a house on good, solid foundations or on rocky foundations? Creating from a place of joy, of peace, of harmony with yourself, and with the intention of well-being for all, will give birth to a creation that embodies that intention. Creating from joy means creating with that part of you that remembers what it is like to play, to laugh, to be a true co-creator. So, re-engage with your inner child and go have some fun! Even if you can't remember how to play anymore, fake it till you make it! Read an imaginary story, do some cartwheels, roll down a hill! Life is meant to be *fun*!

CREATING FOR HARMONY

The natural world (that we are a part of) already exists in perfect harmony and balance (at the level of unity consciousness[87]). So, wouldn't it make sense—if we want to create perfect harmony and balance in the physical reality—we should create from the place where it already exists?!

The Indigenous talk about creating for Seven Generations. If you think about it, it's an incredibly difficult proposition, especially from a 3D mind. It means that we need to design and build systems today that will still work in harmony with all things and continue that way for at least 500 years. Think about the level of complexity that would be required! We build buildings and infrastructure now that are not in harmony with, well anything, for even *one* generation.

We would need to be able to consider all things and design/build from that place. For example, migratory paths

---

[87] The ONE I AM, that our higher selves are always tapped into.

of wildlife, calving or spawning areas, hibernation dens, medicines or tree communities that are working together to sustain the ecosystem, the soils and microbes, the vertebrates and invertebrates, the plants and the birds that call the area home—all working together to sustain life on this planet for all beings. Now, wouldn't that be a fun puzzle to try!

Well, from our lower minds, that's pretty tough. Imagine the trillions of variables that there are to consider. We consider maybe half a dozen variables at best these days. The only way to be able to truly achieve this is to create from the high-mind—from the BIG mind. Tap into the motherboard, awake to our connection with the land, the Earth, and all life. Feel it and create *for* harmony, *from* harmony. Then the connections will reveal themselves and we can truly roll up our sleeves and have some fun!

PRACTICE BY FOLLOWING YOUR INTUITION

It's that quiet voice inside you. It's your higher self, and it knows what your purpose is, where you are going, and what lessons you will learn along the way. Not a bad voice to listen to! It gets stronger and louder when you start to listen to it and apply its guidance (when you are ignored long enough, you stop trying to be heard, unfortunately). It is the wisdom of the All-Knowing that is flowing through you. Aren't you kinda curious about what it's got to say? Start to trust yourself. You know what is best for you!

I once spent an entire day with no plan, just following my intuition from place to place. It was incredible the places I took myself and the people I met! I would stand on the sidewalk and say to the Universe, "Okay, what's next?" and then follow my intuition wherever it led me. I did look a bit strange standing there waiting for instructions from an "unseen" source, but it was incredible! I ended up getting my whole "to do" list for the month done in one afternoon, and in such creative and unimaginable ways my low mind would not have ever thought possible. And I had great fun doing it—the people I met and the places I visited were so cool. Because the high mind is limitless—meaning the possibilities are limitless, and therefore also way more creative and efficient—connecting ideas and people in ways I would never have thought or known to do without this guidance. To this day, it was one of my most productive and incredibly rewarding days.

But don't get discouraged: strengthening your intuition takes time. I mean, we have ignored it for a very long time, so it's got a bit of stage fright! Be patient and keep at it. Look for the small miracles that show you it's working. They are all around you and the more you look, the more you will see. Because guess what? Your true self really wants the best for you and really wants to be your best friend!

SET YOUR DEFAULT TO THE POSITIVE

"Think happy thoughts and you will be a happy person". We have all heard this before. But there is now scientific study that show your cells learn, adapt and grow, so be careful what you are teaching them!

Negativity distorts and positivity uplifts and brings back to wholeness. The problem is, most of our thoughts are going on in our subconscious, so there is a bit of work involved in retraining our minds. Creating a default setting to the positive is basically creating a habit to the positive.[88] Make it your go-to reaction. When you catch yourself spiraling, immediately find the opposite aspect. Everything is aspects of harmony. It is said that every 'negative' has

---

[88] Or what we *perceive* as the 'positive' as there really is no 'negative' and 'positive', just different experiences and lessons along the way. Try for a day to see them from a neutral mind and just observe. Become an observer of your mind. It is a fascinating experience.

a corresponding 'positive', and vice versa. And the one you feed gets bigger and stronger.

> Love – Hate
> Compassion – Disdain
> Hope – Despair
> Acceptance – Anger
> Trust – Deceit
> Confidence – Lack of
> Faith – Doubt
> Ego – Withdrawal
> Belief – Void
> Joy – Anxiety
> Life – Death
> Dreams – Nothingness
> Motivation – Complacency[89]

I had a very hard time going from zero to a hundred. It is not easy to break ingrained patterns and habits of negative thinking, self-judgment and worry. We do not need to be hard on ourselves and expect to go the whole distance at once. The amazing thing is, all we need to do is *choose a little higher* in every action and interaction.

If you feel anxiety (low frequency), instead of reaching for a bag of chips (low frequency), muster up all your strength to reach for that apple or even a bag of organic popcorn is a higher frequency! Or, if you feel self-doubt, try to have just a drop of faith. Just one little drop. That simple action will automatically raise your frequency and you will start to feel better. Then choose a little higher with your *next* action and so on. Before you know it, that funk you were in, is a thing of the past!

THINK BETTER TO CREATE BETTER

What you see is a manifestation of what was in your mind previously. Everything was energy (*a.k.a.* a mental

---

[89] The Transcendors, by Rik Thurston.

image)[90] before it became form or physical manifestation. Think well; you will create well. Think beautifully; you will create beauty. Challenge yourself to think the way you wish to create and not just what society tells you is what you should want. Strive higher! Use nature as your barometer because you had a hand in creating vast oceans and mountains, beautiful flowers and complex organisms. So, don't wimp out now!

When you are just getting started, have a go-to image in your mind funnel that immediately makes you feel happy, elevated and peaceful. (Mine is a beautiful pink flower with a yellow center and lush green leaves.) The more you can feel it, describe its lines and shapes and tones, its smell, its vibrant color, the more your mind is now holding a focus.

Focusing on anything for a while is a meditation which will pull you out of your negative thoughts. The more you can give the image detail, the easier it will be to hold the visual image. Just be patient and keep practicing the positive until you can find your own balanced state.[91] Your mind is the creator of your reality! How you think is how you will be, over time.

But don't get discouraged if you don't see it manifesting right before your eyes. It takes time and there is a lag period in the universal energy field of at least a few days and up to months or even years depending on the size of what you are manifesting and how you have put out the intention. A tree will not grow overnight. It takes time and patience but then it is there.

The young generation today is being born with superpowers well beyond the older generation. The world may

---

90 Well, everything is energy, so everything is created first in the energetic realm and then, by the time we see it in physical form, it is technically already in the past.

91 This works wonders if you are starting to feel a bit stuffy, nose running, under the weather. Change your thoughts, put a nice watercolor in your mind funnel, and you'll be on the road to feeling great again!

not make sense to you as it is at the moment. I agree—we are all waking up from a party gone wrong. But don't give up! It is coming and you are right!

WE NEED YOU TO CREATE THE NEW WORLD WE ALL WANT TO LIVE IN!

So, let us all stop focusing on the problems—dying whales and the trees and pollution in the water. Let's not *ignore* the problems. Be *aware* of them and then actively work to improve yourself and the world we live in. But the most important and powerful way you can do that is to focus your energy and emotion on what you *do want to see* rather than on what you *don't want to see*.

Remember, we are powerful co-creators. We manifest whether we like it or not, and whether we realize it or not. By focusing our energy and emotion on the problems of the world, guess what?—we create and perpetuate more of the same energy, which then manifests into more of the same in physical form.

Mother Theresa once said that she would never go to an "anti-war rally", only to a "peace protest". She knew that all that energy focusing on 'not having war', was still focusing on 'war'. Whereas, what is it we really want? Peace! Clean oceans! Beautiful, healthy forests! Healthy nutritious food in abundance! Feelings of joy! Smiling faces, caring and compassion from colleagues. Wonder and curiosity, excitement and creativity!

It's not hard. Just focus on what you *want* to create, and your mind will start to work at it, and the energy will start to flow towards that eventuality. We are so much more powerful than we realize. And imagine if we all focused on clean and healthy oceans, forests, food, compassionate relationships? Wow, that would be incredibly powerful, directed, manifesting energy! What could we create?!

Stop right now and create your own default setting: one image. An orange, a strawberry, a beautiful flower. Give it a texture and a color, maybe even a smell. Keep it simple. Now, whenever you catch yourself spiraling or getting down, grab this image and focus on it for a few seconds or a few minutes if you can. It brings your mind back to you and under your own control. Now you can direct it to where you want it to go. A nice walk in the forest, a warm bath, writing a poem, painting a painting, dancing wildly to your favorite song, or sitting with your breath. You have the power to create how you feel!

BRING THE FUTURE INTO THE PRESENT

When we hold a construct of where we want to be in the future, if we continue to hold it as a future construct, guess what?! It will always be a future construct! That's why I always had a hard time with "the healing journey". Yes, it is a journey, but if we always treat it as a future destination, we will never get there. So, bring that vision of yourself—healthy, vibrant, thriving, happy, strong, connected, your true self—into the present. Talk about it as if it is already here. *"I am healthy today. I am strong today. I am my true self today"!*

I went to a party once where the theme was "Come as you will be". We all had to act and converse as the future version of ourselves—one year from the current date—talking about how our projects have come to fruition, how our relationships are thriving, how we have great new clients, etc. It was great fun! And there is no more powerful way to manifest than to believe it is already here.

INVITATION

"I am happy, healthy, loved".

"I am living my true purpose, thriving with joy and filled with love".

"I am prosperous and create abundance in all I do".

"I am in a relationship with myself that is sensational"!

"I am a great speaker, writer.... [fill in the blank!] and fulfilled in all my work"!

TUNE UP! (*A.K.A.* ENERGY HYGIENE)

Clear out some space in the noisy food court of your mind. Our breath is one of our most powerful tools, if you learn to use it properly (mediation, belly breathing... yes, all that good stuff). It is highly complex technology that was built into the design. And in order to receive that golden guidance from your higher self or the Universal Mind, with all that is being thrown at us, the most important thing we can start with is breathing to open space in our minds.

Our minds are so noisy! Waayyyy too much going on—noise, clatter, thoughts, plus all the external stimulants we are throwing at it constantly. I should know: my mind can at times be literally like a food court at lunch hour! We need space to receive and quiet to hear. How are we ever going to receive or hear guidance and connect to the "motherboard" to create an incredible future or even just to find internal peace?!

If you notice, nature and Source intelligence creates the most brilliant designs with immaculate and highly tuned precision. Yet it is always simple. We have way-overcomplicated our lives and think therefore that the solutions to health, happiness, and prosperity need to be complicated. So often we get bored with the simple brilliance that we are built with to get us there. Life was not meant to be a struggle. Breathe to create space; breathe through uncomfortable emotions to dissipate; breathe through energy blockages or dis-*ease* to heal. What you *resist*, persists. What you *breathe through*, releases and eventually clears. There is nothing that the breath cannot get you through.

So, clean out that gunk; get into all those crevices and clear it out![92] Sit in stillness and breathe through all the noise to create a clear pathway. We can only hear our intuition, guidance, answers, and wisdom when we create an open road and some silence in our minds to hear it. Manifesting requires space to create something new. It's like a new relationship. You need to make space for it in your life or it just won't work out.

This is energy hygiene. Remember the part about us being energy beings? Well, just like our minds need to be cleaned up regularly through meditation and breath work, and our physical bodies need a shower from time to time (soap is always a good thing!), our energy body also needs cleaning up along the way of life's adventures or it gets all dirty with other peoples' energies, frustrations, anger, noise and all sorts of gunky build-up. Ewww. And it drags us down and makes us feel really tired, irritable, moody and even depressed.

So, if you are feeling down or depressed… or, all of a sudden, a dark mood comes over you and you don't know where that came from… give your body a shake, chant a mantra, or

---

92 Remember, your mind creates your reality, so getting your mind cleaned up and focused is the most important cleaning you can do for your "Ferrari".

put on some high-vibrational music for a few minutes. You probably picked up an energy bug from someone having a bad day. It'll clean that right up! But remember, going into a zone of contamination, put boundaries around yourself and your energy field. Prevention is half the battle!

For me, cleaning up my energy body is a daily practice and just takes a few minutes before bed so I have a good sleep and I do not have to sleep with all the other people's energies I encountered that day (I like my bed to myself, thank you!). Sometimes I spend more time on weekends. Cleaning up your energy body is just like tidying up around your home all week to manage the mess and then getting to the floors, windows and tight corners on the weekend.

I write in a journal whatever comes and when nothing comes, I scribble random words and thoughts until I get the flow. I swim to clear the energy and negativity of the day, in natural water when possible. I go for a barefoot walk on the grass. I meditate daily and throughout the day even a few minutes of belly-breathing does the trick. Sitting in stillness on a park bench for five minutes and just breathing. Focus on a tree or a plant or even a crack in the sidewalk. Focus your mind and breathe deeply.

And then the best part! I have created imaginary energy scrubbers that look a lot like the automatic circular vacuum cleaners, but these guys have hard hats and big smiles. They go in circles up and down my body and use bright, white light to suction and clean. They do a great job—up one side and then the other—sucking up the gunk from the day or even after a tough conversation or interaction. Up and down my spine is the best! It feels great getting into all the crevices. Our mind is the creator of reality, so use your imagination and clean away.

Another important way to keep your energy clean is to eat nutritious food full of sun-life-force energy. Yes, this is also energy hygiene! Because energy cannot be created or destroyed, just transmuted. And guess how we get a great deal of sun energy? Through our food.[93] So, buy local, local, local! The most local you can get is foraging, right in your backyard. I made a stir-fry once with a burdock root pulled from the forest behind my friend's house and some dandelion leaves. My body was literally buzzing with the life-force energy all night long. Pretty amazing what is available to us if we just look. There is abundance everywhere, even in the city!

Bless your food with gratitude[94]—thank the farmers who harvested the vegetables; thank the chickens for giving their lives; thank the sun for giving your food nourishing energy; thank the soil and the wind and the rain. Make it from your heart. This is a true blessing. And by putting gratitude into your food, you begin to clear any negative or toxic energy that may be in your food from all the problems in our supply chain today. With enough practice, you will become a powerful transmuter of energy, bringing light where there is darkness, health where there is sickness. So, start tuning your food—it's what fuels your body. And the gratitude you give will come back in abundance.

DANCE! SING! DRUM!

Another very powerful and fun way to raise our frequency and keep our energy clean is to literally tune yourself! As energy beings, everything is *vibration*, *frequency* and *sound*, as Tesla said. Music will uplift you, but make sure to find the right energy that you want to be tuned to (like a radio station). Mozart has been recorded as extremely high-frequency because he composed his music from a higher state of being. Really good Indigenous drumming and singing is interestingly tuned to the rhythm of nature,

---

93 Well, we used to anyway, but now the weeks on a truck leaves our food a bit lacking for life-force.

94 Another high-vibrational word and action.

which is a very high frequency.[95] Tom Kenyon is beautiful music to meditate to. Binaural beats balance the two hemispheres of the brain.

Look for the frequency (Hz)[96] that is higher than what you currently feel. You will need to feel this and tap into yourself and feel what the music is doing to you. Is it uplifting you or allowing you to wallow in self-pity? If you are still on the couch with your bag of potato chips, feeling down at the end of the song, it's not the right song! Tune yourself until you feel your rhythm elevate to a higher frequency. Don't go to match what you are currently in (unless you are the Dalai Lama).

Words are energy, too! (Do you see a theme here? Everything is energy!) So, the genius of mantras—which are simply words that are chosen in certain meditative practices[97]—is that these words are specifically chosen for their high-frequency properties. Not only do they brilliantly keep the mind occupied so that you can focus inward to meditate, but they also tune you up as you vibrate the sounds. You literally become the musical instrument making the vibration. This is even better than listening to music!

Here are some great ones you can simply repeat or find online and sing along. I sing them when I'm driving, walking to work, or even making dinner. All of them are high-frequency words and sounds which tune me right up and up and up!

Sat Nam, Sat Nam, Sat Nam, Sat Nam, Sat Nam, Sat Nam, Sat Nam, Sat Nam, Sat Nam[98]

---

[95] The Transcendors, through Rik Thurston, said each soul grouping has a different frequency. It is said that the Indigenous soul grouping came down tuned to nature.

[96] Each note or vibrational sound has a frequency attached, as do our bodies, our moods and diseases. Always tune *up*, not down!

[97] Kundalini yoga uses mantras as well as Transcendental Meditation, made famous by the Beatles.

[98] Translates to "I AM"—one with Source and all that is, and that is my TRUTH'.

Guroo Guroo Wha-hay Guroo, Guroo Ram Das Guroo[99]

And pretty much anything by Snatum Kaur. Her music was a true blessing for me during my tough moments. Her beautiful angelic voice and sounds were always able to help me lift my soul and spirits. I must have listened to her songs on repeat thousands of times, until the vibration literally lifted my heart. This is one of my favorites. Jump to 3:40 minutes in for the chorus or chant the whole song as a full meditation!

If you are feeling particularly frustrated or angry, belt it out! Belt out any song as loud as you can! Ever hear a really incredible Indigenous drummer and singer at a pow-wow? The way they sing from the depths of their souls, is the most powerful thing you can experience. Here is the song by Buffy St. Marie from her album *Medicine Songs* that I sing as loud as I can when I'm stressing and just need to get it out.

Choose whatever keeps your energy clean, tuned up and flowing. And remember, it's all ultimately powered by our thoughts, so control your thoughts and you will control

---

99 This is a mantra of self-healing, humility, relaxation, protective grace, and emotional relief.

how you feel and your vibration/frequency. You could be feeling great and then allow a negative thought or experience into your mind and, all of a sudden, it changes your mood. Thoughts are energy. Period.

When all else fails, simply lie down on the Earth and ask our Mother to help you heal... to take away your emotional pains and sadness. She is our Mother Earth and she will embrace you. The Earth can literally suck up all that gunk from inside of us, she is that powerful. So, lie down on the Earth with your whole body pressed up against her, close your eyes, and ask for support and healing. She is there for you.

ELDER WHABAGOON BLESSING:

Pick a morning, any morning, to lie down on Mother Earth. Feel the morning dew/water between your toes as your feet connect with the Land. Feel that connection through your mind, body and spirit. Lie down and feel your back connect with the Earth's vibration. As you lie there, take in a deep breath while remembering Tree Nation provides the oxygen you breathe in with every breath. Look up into the Sky Nation and take in the stark blueness of the sky. Watch the winged-one fly/soar and appreciate by acknowledging you as Creation.

Take care of this Sacred Land we call Mother. Be one with our Mother Earth. Pause, Be ONE with her. She loves you.

BREATHING LIGHT MEDITATION

Close your eyes. Take five long, deep, slow breaths. Feel your lungs fill up. Breathe into your belly and allow your belly, rib cage and lower back to expand on the inhale. Breathe in for two to four counts (whatever feels natural to you where you are at) and out for double that (four to eight counts). Focus on the point between your eyebrows with your eyes, still closed (in your mind's eye). This is your third eye. Sit in stillness and just breathe.

Imagine, *on the inhale*, bright emerald-green energy is coming up from the center of the Earth and flowing up through the souls of your feet... up your legs and hips... and begins to swirl in your abdomen. Then imagine bright white light is coming from the Universe in through the top of your head... and filling all the corners of your body... from your head to your face... neck... shoulders... torso... back... heart... and then begins to swirl in your abdomen with the great Earth energy... creating a beautiful vibrant color swirl. Imagine all your cells sparkling with bright light... and your veins coursing with this beautiful green and white light.

Now imagine, *on the exhale*, the light is spilling out through your fingertips... and the bottom of your feet... spilling out all around you... to form a light that surrounds you. It radiates out to the walls in the room... to the whole house/apartment you are in... to the city... the country... the Earth... the Universe.

This is who you *are*. This is the light within you that will create the change in the world. Sit in this light, feel it, radiate it out and share it with the world every day, as many times as you can. When you walk down the street, imagine radiating this light and love from your heart to each person you pass. Send them love and light. Send them some of this beautiful light that you have within you. Share it with as many people as you can.

ZIP UP TO KEEP UP

Now that you have created all that vibrant, positive and bright light energy in and around you, zip up your energy body to define it as yours. It is a crazy place out there: when there is chaos and confusion, bright lights like you are constantly at risk of being drained. No, this doesn't mean to stop sharing your light. It means to put up secure boundaries while you share.

By zipping up with an imaginary zipper each of your chakras—from your root chakra to your crown chakra and beyond—and then putting an imaginary boundary of purple lights, pink roses or a silver shield around you, you keep your energy boundary clear and make clear that you don't take on others' negative junk. The great thing about these boundaries is that they only allow love (*a.k.a.* the highest frequency) in and out!

So, you can still radiate your light to the world, shine bright, and receive light and great energy from others. But no one can take your light and energy away without *your permission*. There are real energy vampires out there.

Yes, they do exist! They suck your energy dry, often not consciously even knowing they are doing it. So, don't get angry or blame them. Simply make the choice to not let others take away what you work so hard to build up, and do this with strong boundaries. Draw them out with your mind. Your imagination is powerful. It is the tool of co-creation.

As we go into greater uncertainty and fear all around, stand proud in your light and know that you are leading by example. No, others cannot suck you dry, but they can ask you how you stay high and what they can do to vibrate the same way. Then you can help others on their journey to higher frequencies and dimensions that are part of the infinite possibilities we are now aware of and part of creating.

The best part is, when you stand in your light, nothing can enter your energy without your permission. Remember this. Know this to be true. If you are going into a grocery store or work or anywhere you may feel energy (physical or otherwise) could come at you, set your intention, zip up, and put up your boundaries. And then go out in the world as the Heart-Centered Warriors that you are, without fear and without trepidation. You've got this!

GROW YOUR ROOTS

We are all in some major need of grounding. I know it is my daily challenge. A tree with strong, deep roots can withstand large winds and storms. Grow your roots from your feet to the center of the Earth to be strong and grounded no matter where you go.

Imagine that there is energy coming from your heart down through your root chakra... through your feet... up through the top of your head... through your armpits... through the tips of your fingers. Imagine you are a BANYAN TREE with its massive root system (if you haven't ever seen one, look it up—it is spectacular!)... strong roots reaching out and grounding to the Earth... to the energy field that you are a part of... to the energy that exists all around you. Anchor yourself and you will not be swept away by the chaos all around you.

When I'm out in nature, I ask the trees to show me how! It's a great ice breaker!

SMILE!

With a smile, your body instantly begins to relax. It is the simplest, fastest way to begin to bring calm into your body. From there you can begin to see things just a little bit differently or just enjoy a moment of peace. So, press the auto-relax button on your fancy vehicle, and smile.

A kundalini yoga master teacher and friend of mine developed a meditation that he calls the "Smiling Buddha". It is really simple: sit down, cross your legs, put both hands up, make the first two fingers into the "peace sign", and then close them so they are touching. Close your eyes and smile.

That's it! For eleven minutes, or two minutes, or as long as you can![100]

Smile! You are beautiful!

---

[100] Sat Tara Singh Khalsa, Mexico

FORGIVE YOURSELF

Brace yourself, this is a big one! We all need to embrace forgiveness. We all need a massive Ho'oponopono, an ancient Hawaiian practice of reconciliation and forgiveness. By repeating these words, you send out forgiveness to all you may have harmed in this life and in past ones; and you ask for forgiveness from all that may have harmed you in this life or prior. It is one massive global clearing of Karma. It doesn't matter the specifics of who, what, when, where. Just Ho'oponopono and clear it all away! It feels great—give it a try!

"I love you, I'm sorry, Please forgive me, Thank-you".

There is a major reset coming—all the old is being brought to the surface to be healed and cleared. We are shifting as a humanity. So, let go of all the shame, guilt, anger and sadness. And definitely don't hold onto any self-hatred. Just not worth it!

FORGIVE YOURSELF! Let it all go. We have all done things we are embarrassed about or feel bad about. Make amends with yourself. Holding onto that is like holding onto poison. We must put all that baggage down.

Imagine filling an air balloon full of all your baggage... all that resentment... anger... fear... disappointment... guilt... and shame. Put it all in there!

Then cut the ropes and watch it fly away into the ethers. Watch it get disintegrated as it crosses the Earth's atmosphere. It is gone.

Forgive yourself. Let it go.

PRACTICE RECIPROCITY: LET THE ENERGY FLOW

Okay, so you may have noticed a theme here: energy. Yep. EVERYTHING is energy.

Think about how to keep that energy *flowing*. Energy wants to move, to jive, to rock and roll! Nothing in nature is static. The grasses move in the wind; the water flows down the rocks; the ants build their homes; the hawks glide beneath the summer sun. It is a symphony of the highest order, and it is happening all around us all the time.

A friend of mine, an incredibly brilliant North American Indian leader, taught me that the term "Indian giver" was originally given to the First Peoples because they would give blankets to the Europeans in the winter when it was cold and then take them back in the spring when they were no longer needed, to be used for another purpose.[101] Everything was always in continuous motion. It wasn't: *This is mine and I hold it for myself, even if I no longer need it or it's not being used.* It was: *What is needed by whom, when, and then how can it be used for another purpose?*

The future is a sharing economy—not just physically exchanging goods but also *energy* exchange. Why? Because it feels good and that is how natural law works. Everything is energy. Let it flow! It will come back to you in abundance. And that is how we create abundance for all.

I always marvel at marinas on a nice sunny summer day when the wind is just right—perfect for a beautiful day sailing on the lake. Yet the marinas are always full of sailboats. Isn't it amazing? We kill ourselves working to buy things that we think we want but then don't use and have to work even harder to maintain and keep. All the while we dig up resources, shipping them across the world with fossil fuels, all in order to sit in a marina barely used. Does this make sense? Is there not a better way we can come up with, with just a bit of thought and creativity?

Movement and use keep things from going stagnant. In our bodies, stagnant energy manifests as cancer and other forms of dis-*ease*. And letting things around you in your life stagnate, creates stagnant energy in you, too!

Ever notice how things just tend to fall apart if not used for a while? The more you can use your things, interact with the energy, share, pass on, re-use or re-purpose, it will create a vibrancy and elevated energy of all those things, and in return, in your life.

Life (energy) wants to be interacted with, used, played with, engaged with. It is not there to be stared at or to sit in a storage container. The more it is used and interacted with, the more life energy is put into it and the more it

---

[101] Karl Shay is an American Indian Knowledge Keeper from the Penobscot Nation, and a Mikmaw of the Lnu First Peoples, Keepers of the Eastern Door.

transforms the space it is in. This is how we create environments that are teaming with life and abundance. So, yes, do it (the sharing economy) because it is good for the environment, because it feels good, because it builds community. And know that by doing it, you are uplifting yourself and keeping the energies high for all.

In nature, this reciprocity is seen in how all life cooperates. The trees support the birds, the roots support the microbes, and all are supported by the rain and the sun in return. Reciprocity is not a two-way channel but a multi-dimensional relationship that builds community. A tree will not give life to a plant and then want something from that plant in return. It gives what it has to give, knowing it will get what it needs when it needs it.

Charles Eisenstein, in his book *Sacred Economics* talks about how "the gift moves towards the empty space". In ancient cultures, people were taught to give what they had to give and trust that when they were in need, someone else's gift would find them. What you give will circulate until it finds the void where that need can best be fulfilled with your gift. In this way, a cycle is established where all needs are always fulfilled.

How do we make this philosophy work for seven billion people? How do we learn to trust one another again, to break down those walls and silos of separation, and remember that we are all one and we are all in this together? Can we collectively intend to open ourselves to this higher realm of understanding? Can we take the leap?

To give to another in need is a truly beautiful thing. It is more of a gift to the giver than to the receiver. It is not a zero-sum game out there, folks—unless you choose to create that reality for yourself. If you choose to trust in the Universe and in natural law, each person will be taken care of.

The challenge we face is that currently we exist within a system founded on no trust, on stagnant energy, on hoarding, on fear of letting go, fear of sharing, and fear of giving away. Natural law will always rule the day, but it may take some time and require some patience to stick with it before we get back to that place!

Elder Shelley Charles once said to me, "It took effort and personal responsibility to make community work. You had to work at cooperation, develop your own skills and contribute to the whole. It didn't just happen".[102] We have abdicated responsibility to external sources to make *them* work for us rather than making cooperation work for us all.

Life is most rewarding when it is lived in service to others. Build into your day to add value and do something nice for one person. It can be opening a door, smiling at a stranger, giving a homeless person your pocket change or buying them a warm tea. Everyone (including each one of us) has good days and bad days. Every single person (including ourselves) needs a helping hand from time to time. Be the one to offer that hand when you pass another human relative in need. Then when you are in need, the Universe will bring *you* a helping hand.

But don't do it to expect anything back because that creates an energy of wanting something in return, which is just lower frequency energy. Give to give, as if you will never receive anything back from that person because, chances are, you won't. When your time of need comes, and it will, that stranger you held the door for will be long gone. But there will be another person to hold the door for you.

This week commit to giving away one thing that you no longer need, whether it be clothing, money or food. Give it away to someone who may need it more than you without expecting anything in return. Give it without attachment for a return. Conversely, if you receive something from someone as a gift of any kind, big or small, look them in the eyes and say "thank you". Receive it with gratitude

---

[102] Elder Shelley Charles of the Chippewas of Georgina Island First Nation.

and appreciation. This honors the giver and gives them the gift of giving.

Now try this with the food you eat. Thank Mother Earth, the sun, the rain, the soil, the microbes, the farmer that tilled the soil and planted the vegetables, harvested and cared for them, the person who prepared the meal, and all the energy that went into it.

The future is reciprocal relationships with all things. And this is a gift to us of infinite abundance.

THE AGE OF CELEBRITY IS OVER. SING YOUR BEAUTIFUL SONG!

We look up to celebrities in many ways because they are living their true essence whether it is an exceptional athlete, an incredible singer, or passionate actor. We all have that ability to live our true essence. Each of us was born with our own inner genius and no one is greater than another. In fact, each and every one is necessary for the whole to truly thrive. We are *all* celebrities. So, sing your beautiful song! Right now! What is *your* inner genius? We all have it!

SING YOUR BEAUTIFUL SONG!

Now's the time. Don't hold back. Let's hear it. Sing as loud as you can, whatever it may be. Tell your jokes, write your blog, draw your heart out, or if you don't know what your "thing" is, just smile at another person, and stand confidently in the knowing that:

> *YOU ARE AN AMAZING INCREDIBLE UNIQUE INDIVIDUAL THAT THE WORLD DESPERATELY NEEDS TO BE YOU! NO ONE ELSE CAN DO IT! JUST BE YOU AND BE PROUD. WE ARE NOT OUR BODY. WE ARE NOT OUR FORM. WE ARE ALL POWERFUL SOULS WITH INFINITE WISDOM. WE ARE ALL SUPERHEREOS. YOU KNOW IT IS TRUE!*

For a long time, I had no idea what I was good at, what made me tick, what my inner genius was, or even what I enjoyed doing. And that would make me super sad. It is an explorer's journey. Set out to explore yourself. Go out on dates with yourself. Try different things. Play around with what excites you and especially with what makes you nervous!

I was terrified of public speaking and so I signed up for Toastmasters. I went every week until I was winning competitions. Now I love speaking in public, but the first six months doing it, I was petrified. That took a lot of courage and willingness to be honest with myself—to know that we aren't going to be great at everything right away, but that you can be great at anything if you put your heart into it!

BEAUTY IS DIVERSITY.

Nature builds with diversity. No two flowers, plants or trees are the same. Yet each has its purpose in making the ecosystem thrive. That is why we are so diverse and that is so important to maintaining a healthy whole.

If we were all like our favorite celeb, it would be like a Christmas tree farm. The trees grow to die and not much else is going on there. *Reconnect with who you are and live from your true essence rather than following others*. Sure, we can admire those who are standing in who they are, but use it as inspiration to then stand in who YOU ARE. Acknowledge their light and then get on with being YOU!

I want the Earth shaking with the beat of the drums, the air thickening with the smell of sage and sweetgrass, feet dancing to the rhythms of beauty, and the birds singing in chorus to our beautiful songs! Each person has a song to sing. Sing *your* beautiful song.

That is your contribution to this world. That is your purpose! To be YOU!

COMMIT. DISCIPLINE MAKES SUPERHEROES OUT OF EVERYONE

Make healthy choices from your high mind and commit to them. Decide to meditate each morning for a few minutes? Commit to doing it.[103] Ride your bike to work? Commit and do it. Eat healthy for a week? Commit and do it. *Just do it*, as the saying goes. You will feel such a sense of accomplishment from training your mind to do what you set out to do.

Start small, yet be consistent. Small things are great to build good habits and a strong and resilient mind. Start where you are! Make your bed each morning. Commit and do

---

[103] Everyone should meditate daily, especially in these crazy times. That, in my opinion, is a must! If you haven't tried it, get out there on your seat cushion and breathe. If you are too intimidated, go take a meditation course. There are tons out there! It is the greatest gift you can give to you.

it. Nature is always moving. It doesn't get lazy and just say, "Nah, queen bee, not really feeling the vibe out there today. I'm just going to stay in the hive and chill out today, drink some nectar and veg out". They go out and do their thing.

We are so used to things being done *for* us. Fast food. Fast energy through caffeine (rather than meditation which will give the same but longer lasting boost!). Immediate "joy" through shopping. Entertainment by pressing a button. We are used to being able to buy "happiness", "health", "well-being".[104]

Traditional cultures (and our ancestors) used to spend time learning art and music, gathering in costume, and carving their own instruments in order to entertain. The *process* was part of the entertainment. It brought joy to people as it gave purpose and skill, a contribution to the whole.

We have outsourced our own value and our own ability to contribute meaningfully to the whole. The making of a tasty jam is no longer the contribution of a skilled grandmother, rather a factory. The food, the equipment we use, the things we buy are no longer someone's special gift contributed to the community. It is at the cheapest price, sent from far away, made by someone under questionable rates and circumstances. Then we buy it only to throw it away or rarely use more than a handful of times.

My business partner Jerry's daughter, Geraldine Asp, once gifted me a beautiful hand-made bag during a visit. She said to me as she handed me the bag, "I felt you could use something made by hand with love". It is a simple bag and yet the one that I somehow end up always using. It just feels *right*. We have traded the value of love, community, friendship, generosity and contribution for fast, easy, convenient—quantity over quality. I think *we* got the short end of the stick.

---

[104] If our idea of health, happiness and well-being is fleeting and short-lived, like a seagull who just scored a beaten-up French fry.

Committing to quality over quantity takes discipline. It takes owning your responsibility at building your skills and offering them to others in service. The truth is it takes a huge amount of discipline to be your true self *right now*. There will come a time when we have succeeded at creating a world that works for everyone and then it will be easy. Yes, for now we have to work it because we have created a world that does not currently support our greatest potential. But you will still be rewarded by being your true self because is there anything better? You are awesome, amazing and literally 1 in 7.8 billion! So, stand proud in who you are and work it!

Your higher self knows where you are going and what you want to achieve. So, let go and trust that you will go there and do all that you need to along the way to succeed and prosper in your life's mission. Which, by the way, leads to an abundance *in* life—true abundance, which is much more comprehensive and encompasses so much more than material possessions. It is intangible wealth over tangible wealth.

Intangible wealth is priceless because the material realm is just one small aspect of who we are. And that is so last year! True prosperity is having a full life. What do you want people to say at your funeral about you—that you had a bunch of cool toys *or* that you were a great person full of kindness, generosity, humor and love, who brought joy to all those around you?

> *"We are moving to a reality where we are creating for our soul, not just physical, needs".*
>
> —NIKOLA TESLA

Tell your low mind to take a chill pill. Mine always likes to be the center of my attention with many fantastic reasons why I should buy this or that, eat this food, binge out on TV. It is brilliant: don't get me wrong. My mind has served me well through law school and all my work. But it was just never meant to be driving the bus! It is in unknown territory and we gave it full reign. No wonder it is acting like a teenager out on their first night on the town with a fake ID.

I tell my low mind how amazing it is and ask it to then kindly serve my soul. I let it know that it has a very special job that only it can do. Whenever your mind tries to get in the way with thoughts of self-doubt, put-downs, negativity or the 101 reasons why you should just go for the fries instead of the salad, or why it's just easier to get someone to do something for you than doing it yourself, say "I love you" and keep going! You are a powerful light-being, Heart-Centered Warrior, co-creator of Universes! What are you waiting for?!

GET UP AND MOVE!

Right now, get up and shake, shake, shake! Move, dance, jump, whatever you feel in you to do. Shake off that stagnant energy! Shake off those negative energies from the day! Shake off your feelings of guilt and shame.

Shake, shake, shake! Move, move, move!

Put on a great song and dance without inhibitions. Without caring what people think. Early in the morning, I will close the door to my room, put on a great song, close my eyes and just dance my heart out. It is a great way to get the energy moving and stretch out those dormant muscles! Go for it! Make it count! Who cares what you look like! The "stranger" the moves the better!

We stagnate energy when we sit, and that collects and gets icky. Then we feel icky. So, get that energy moving! If it's not something you want to stick to you, get up and move it out! Nature is the best medicine and the fastest way. Walk in nature, walk the streets, pace in your house. As soon as I start to feel stressed, spiraling, or even craving junk, I get up and go for a walk. Even a 10-minute walk is a great reset.

Walking is an art form and the one I now choose most often. It gets you in touch with your body and with the world around you. My grandfather used to walk ten kilometers a day and, in the winter, would do laps in the basement to get to 10km, well into his 90's. The repetitive, consistent rhythm of it can be quite meditative and really begins to work your whole body. I go for a long walk each morning and it keeps me feeling good.

Whatever your form of movement, just get up and move!

HUMOR

The one thing I have seen transcend all the wise people I have known—from my grandmother who was always making people laugh, to Jerry my business partner (who has the best jokes and greatest sense of humor no matter how tense things can get), to certain high-level shamans I have worked with—is a tremendous sense of humor in life. If you have ever read the book *Book of Joy*[105] (a series of interviews with and experiences of the Dalai Lama and Desmond Tutu together), you will see this as well. It is all humor and joking with each other, making light fun of one another.

Some call it "resiliency", but it is also wisdom. A knowing that, ultimately, we are spirit beings and this is a human experience. And it is supposed to be fun! Work the deal and strive for the best... but with a lightness about you.

Once you know that we are all energy and everything is energy, you become elated with the joy and beauty that is around you. Your heart is uplifted by the big and the

---

[105] Lama, Dalai, et al. (2016). *The Book of Joy*. Penguin.

small. Life becomes fun. Stop and look at the beauty that is around us all the time. Look at the marvelous creation that we are a part of. The clouds, the sky, the grasses, the flowers, the trees, the birds, the animals—even we humans are beautiful creations! Have you ever watched animals in their natural state, they are so playful.

Some of my best experiences are travelling to remote destinations with Canadian Indigenous leaders because we are always laughing. The jokes and the stories, all told so eloquently, entertain and uplift. There was something beautiful about the oral traditions—people learned how to be incredible storytellers, passionate speakers and entertainers, all without turning on the TV. I love these moments of simplicity and beauty. They are an art form and one I cherish greatly.

My grandmother recently passed away, and one of the things that everyone remembered about her was her fabulous sense of humor. In any situation, no matter what, she did not sweat the small stuff and was always encouraging all of us to smile, to laugh, tell jokes and enjoy life.

Great Spirit has the best sense of humor of all. You will begin to notice this as you begin to co-create consciously with Great Spirit. What you seek will never happen how you thought it would but will always make you smile. This is how you know you are working with Great Spirit directly.

BELIEVE IN MAGIC, CREATE MAGIC

Children are powerful creators because they still believe in magic. Magic is real in all forms we think it is. But it isn't magic really; it's our own power to manifest being shown to us through glimpses as we open ourselves to that possibility.

We are all capable of manifesting greatness, connecting with others telepathically, moving energies, even walking on water (yes, we can do that and more!) and communicating with plants and animals. If everything is energy and we are all one, there really is no separation between us and anything or anyone else. Our perceived separation is just an illusion. Our only limitation is our *belief* that we are separate.

I started practicing with parking spots and green lights, food in my garden and in grocery stores. I travel so much that I have also mastered airports and transportation. Gates seem to open up and lines clear just as I'm approaching.

Magic does not have to be something ostentatious like moving a cup across a table with your mind. It's applied spirituality and it's very practical. Sure, I could spend my time trying to turn a frog into a prince or a prince into a frog (how does that go again?!), but really... of what use is that to me? Then I'd need some food for the frog or really hope the prince is a babe! What I focus on is what I want to create and work towards so I can build a better reality for all. And, yes, sometimes that means I need to catch my bus to finish the job!

Now I work on conditioning spaces and manifesting a healthy abundance of food in my garden. At the beginning of the summer last year, I decided to spend the morning nurturing my garden. It was still early in the summer and not much had come up yet. It is a large garden and I spent about four hours weeding, cleaning up, tending and talking to the plants. When I was done, I went for a long walk. When I came back later that day, the garden had blossomed! Where there were just seedlings had grown full heads of lettuce and kale! There were onions and dill and chives galore. We had a full dinner from the garden that night. It was a lesson to me that when you nurture and give, nature gives back in *abundance*.

What can you start to create in your life using your limitless belief and imagination to serve the highest good? Get in touch with your playful nature and believe the impossible, the limitless, and get to work co-creating your magical world that you want to live in!

GET YOUR HANDS DIRTY IN CREATIVITY AND IMAGINATION—THE PORTAL TO YOUR HIGHER SELF

When all else fails and you don't know what to do and think it is just impossible to connect, grab a can of paint and paint something (my paintings are almost always abstracts and I find that the most fun), carve a cool walking stick, or write a poem even if it doesn't rhyme or even make sense! HINT: Start with your non-dominant hand if you are really stuck: it's a trick to unlock your right brain!

My mom has always taught me that creativity is the portal into my soul. From painting with my hands at two years of age to intuitive painting as an adult, I have learned that creativity is the shortcut into your higher self—into your higher intelligence. It takes you from your left brain (all that thinking and logic that's good for basic

planning) and takes you to your right brain, where all the creating happens! So, get lost in creativity and let your heart shine bright!

We also need to get used to using our creativity to start applying that limitless imagination and outside-the-box thinking to visualize the world we want to see. That's what our imagination is for. No, it's not just for kids and artists. It is the most vital part of the manifestation process. If you can't see it, you can't create it. We are only limited by our imagination. We can train our minds, but if we remain in our limited perception of what's possible, we will create limited realities (sort of what the world looks like out there right now).

INVITATION

Anything is possible! Dream of something right now—big or small... Just start to dream and imagine. But make it goooooood! And by good, I mean as limitless as you can get. I'm talking, reaching for the furthest galaxy... and then even further. There is no limit to what we can co-create.

If all of us, as Heart-Centered Warriors, are going to create this incredible thriving world, we will need our imaginations to imagine what that will look like. But if our imaginations are not used, they get weak and lazy. When we get *fed* imagination (i.e., entertainment) rather than use *our* imagination (an audio book or fantasy story, coloring or painting *outside* the lines), we become limited in what we can imagine our future to be. It then starts to look bleak.

Spend time in your imagination. What do the future possibilities look like to you? What does the vibrancy of a thriving planet look like to you? What are the colors? The textures? The animals? The smells? What part do you play in that reality? Whatever gets you into your right brain and into "the flow".

The flow is a state when you stop thinking and are just doing, moving, being. And it feels awesome! Not forced, not sitting still. Get up and move! At the end you should feel like you just had a great adventure, and it was not planned at all. It doesn't have to be complicated.

For me, if the weather is good, I will get on my bicycle and just ride to wherever the afternoon takes me. Through ravines, to the beaches. Then maybe I'll rent a kayak or a standup paddleboard, check out a farmer's market or some consignment stores, stop to smell some beautiful roses that an older Portuguese woman has tended to so meticulously. Maybe I'll stop for a bite at a corner shop or pick up a yummy vegan treat. Usually there is a conversation in there somewhere with a pigeon or a frog.

What do you do that sparks your creativity, your imagination, your sense of wonder? Grab some colorful crayons or markers and start to go wild! (Staying outside the lines is required.)

HEART-CENTERED WARRIORISM—GET OUT THERE AND CREATE!

Now that you are armed with all these great tools to help you shine bright, what next?!

GET OUT THERE! CO-CREATE!!

As Heart-Centered Warriors, our job is not to sit idly by and smell the roses. It is to tune up so we can be super-awesome Heart-Centered Warriors and then get up and create. Take that great energy out there and begin to change the world! Whatever gets you excited.

The passion, excitement and enthusiasm inside you are Great Spirit *speaking* to you. Listen to it. Follow it. Where is it telling you to go? To be? To do? Make sure it is coming from a place of connecting first and then a place of creating from your HEART-CENTER because that is how to tap into your life's purpose and make the world a better place all at the same time.

Nature is efficient after all, and we are nature! There is no greater joy in life than doing what you came here to do and to serve humanity to your greatest potential. It's what we all came here to do. So, get out there—YOU are a Heart-Centered Warrior! No excuses!

106

That potential may be created by simply radiating your light and love at all you come across... or helping others to let go of their pain and trauma... or teaching others the path back to themselves... shining your beautiful light into the world.... helping humanity heal... or empowering others to empower themselves... sharing a smile or a delicious loaf of homemade bread. That is all invaluable work and a gift to the world! Get out there and do something. Create something. Serve in whatever way you are called to serve. As my good friend likes to say, "Add to the Picasso!" What beauty will *you* contribute?

Ask yourself today: Where and what can I do that best allows my true light to shine?

Remember, whatever action you take is only a vehicle for expressing your true essence—for singing your beautiful song, sharing your heart with the world, and serving humanity. Whether you teach, write, coach, bake, deliver mail or wait tables, do it all from your heart and for the betterment of others.

As long as you are within the realms of natural law, it doesn't matter what you do... just sing your beautiful song!

---

106 Buffy St. Marie, Spirit of the Wind.

I was at a friend's biomimicry weekend retreat a while back and we had one session around the campfire at night, sharing with a partner our inner genius. For many, we didn't know what that was, so we were told to share the first thing that popped into our minds about a time when we felt truly alive and at our happiest. I was paired up with a guy that I had just met that weekend. He was a six-foot-six broad-shouldered engineer—a tough, strong-looking guy. He shared with me that what he loved to do more than anything was to bake bread. It brought him so much joy to bake bread for *others*. He shared that he would bake all kinds of bread with nuts and raisins and different types of flour.

His favorite time of day was inviting all the people he worked with into his office to share his bread. He didn't eat the bread; he brought it for *others* to enjoy. It was a time of friendship and building community that he was creating through his gift. Listening to him speak, I knew that those lucky enough to taste his bread were getting more than a meal. They were getting his whole beautiful heart and soulful energy. That was his true gift he was sharing with others. It is a vision of true service.

Have the courage to stand in your passion! Let your love shine through and we will heal each other and the world. That is the act of happiness, health and abundance.

CONSIDER MORE!

Let us begin to look at things and question what is, to open our minds to the interconnectedness of all things, to open minds to see and experience *more*. Let us open ourselves to see the greater depths of life that exist all around us and of which we are part.

Let us begin to live in awe. This will not only allow us to truly live in harmony and balance with all things by understanding the true impact of our actions and *how* we may create for greater harmony, but it will also give us a level of joy, excitement and adventure we never knew was possible. It will take us past the one room of the house that we have been playing in all this time, to experience other rooms, other dimensions within this house we call "home".

We no longer have the *luxury* of looking at the world inside silos or with blinders on. The challenges we are creating in the world are becoming so grand, so compounded on themselves and so interconnected that we can no longer point the finger at one source. It is, in fact, reflections of our own collective, polluted minds. As Einstein said, "We cannot solve our problems with the same thinking that created them".

So, blow open your mind! Question! HOW can we exist better with all things in greater balance and harmony? How can we build new relationships with each other and all things? Open your eyes to see more; open your ears and listen to what nature is telling you; feel the animals and what they convey. Ultimately, it is all love and it is all ONE, of which we too are a part. Nature is reflecting back to us our own actions and our own states of being while at the same time showing us the pure love that exists within each of us, as the pure energy that is the ONE I AM we all are.

Open yourself to your own bright light and support others to do the same. This is the evolution that is being demanded of humanity right now. It is the era of conscious evolution. It is not something external to us, like the industrial or knowledge revolution. This revolution is going *inward*—back to our true selves, to our true essence, on both an individual level *and* a collective level. And it is happening inside each one of us RIGHT NOW.

Remember the ancient Indigenous prophecy about the Eagle and the Condor (shared by many Indigenous cultures worldwide)? It talks about this period of time in human history when we come to the end of the 500-year period of the domination of the energy of the mind—the aggressive, the Eagle—and we enter a new period of 500 years when the heart energy—that of the nurturing, the compassionate, the Condor—would rise and the two would strive to create balance: the Eagle and the Condor flying in the same sky and creating peace and harmony for all of creation.

This is the opportunity, but it is up to us to *create it*—together. If we are to thrive, it will be only together.

What are we waiting for?! Let's co-create a beautiful world that works for everyone and begin to realize our own human potential!

From my heart to yours.

Light to All.

Love to All.

Life to All.

GRATITUDE & ACKNOWLEDGMENTS

To my mother who has been the foundation, the matriarch, the grounding of creativity and love for me. Thank you for always believing in me and teaching me that anything was possible, giving me wings to fly.

To my dad for teaching me to question, think for myself and stand confidently in the world.

To my sister for being a constant source of inspiration and mirror of pure love and light.

To my bubble partner, here and in the next dimension, Gian, for all you have taught me, and for your unwavering love, belief and support in my superpowers.

To my business partner, mentor and friend, Jerry, for believing in my vision and training me to be a true warrior of the natural philosophy.

To my friend Mary for walking with such strength and beauty, and sharing your journey with me to inspire me to share mine.

To my best friend Lianna for your unwavering friendship all these years and always being there as a shoulder to cry on when things got tough.

To my friend Patricia Fortier for all your encouragement, mentorship, words of wisdom, kindness and support.

To my friend Lawrence for unapologetically standing in a better way.

To my GIDT Board—Sean, Mary and Nalaine—for modeling such strength and brilliant leadership. I am honored to work with and learn from you.

To my GIDT Team—Martina, Judy and Amanda, for your tireless work in service of others, for believing that a better world is possible and working your asses off to make it so!

To my friend Manari for reminding me how to speak the language of nature again.

To my friends Becky and Pimastan for giving me the courage to stand in my light.

To Debbie for allowing me to share the teachings of her late husband Rik and the Transcendors.

To Heidi for modeling the grace that we are all working to achieve and for believing in this project so much that together we made it happen.

To Briana for her help getting this message out—you are a testament to the sharp and quick minds, and deep understandings that the young generation have.

To all the communities I have had the privilege to work with and learn from, thank you. I am forever grateful and honored that you opened your homes and hearts to me.

To Elder Whabagoon for walking the walk of the new way and leading the next generation with all of your heart. Thank you for the beautiful energy you have brought into this book.

And last but certainly not least, to Arzu—thank you for your true friendship, inspiration, guidance and love.

Dyakuyu, Miigwetch, Madu, Sulpayki, Seremein, Gracias, Merci, Thank you.

BIBLIOGRAPHY & RESOURCES

Here are some of my favourite books and resources!

Arzu Mountain Spirit, http://www.arzumountainspirit.com, http://www.arzumountainspirit.com/home-wagiya-foundation.

Bach, Richard. (1973). *Jonathan Livingston Seagull: A Story*. Avon.

Chödrön, Prema. (2013). *How to Meditate: A Practical Guide to Making Friends with Your Mind*. Sounds True.

Day, Tanis, PhD. (2008). *The Whole You: Healing and Transformation through Energy Awareness*. A step-by-step guide to exploring your subtle energy fields. iUniverse.

Dyer, Dr. Wayne W. (1989). *You'll See it when You Believe it: The Way to Your Personal Transformation*. Avon.

Eisenstein, Charles. (2011). *Sacred Economics: Money, Gift & Society in the Age of Transition*. North Atlantic Books.

Gawain, Shakti. (1991.) *Living in the Light: A Guide to Personal and Planetary Transformation*. Bantam.

Goorjian, M. (Director). (2009). *The Shift*. [Film]. Lyceum Films.

Gurmukh, K. K. (2001). *The 8 Human Talents: Restore the Balance and Serenity within you with Kundalini Yoga*. Harper.

Hay, Louise L. (1991). *The Power is Within You*. Hay House.

IISAAK-OLAM Foundation: www.iisaakolam.ca.

James, Kevin. (2013). Humee Hum Brahm Hum. Life is Perfect. [CD].

Jampolsky, Gerald G., M.D. (2011). Love is Letting Go of Fear. Celestial Lights.

Jowett, Geoffrey. (2014). The Power of I Am: Aligning the Chakras of Consciousness. Studio City, CA: Divine Arts.

Kahn, Matt: www.mattkahn.org.

Kaur & Singh. (2009). "Servant of Peace". The Essential Snatum Kaur.

Kaur, Sirgun & Singh, Sat Darshan. (2011). "Bliss (I Am the Light of My Soul)". The Music Within.

Kimmerer, Robin Wall. (2013). *Braiding Sweetgrass: Indigenous wisdom, scientific knowledge and the teachings of plants*. Milkweed Editions.

Lama, Dalai, et al. (2016). *The Book of Joy*. Penguin.

Mipham, Sakyong. (2003). *Turning the Mind into an Ally*. Riverhead Books.

Rong, J. (2004). *Wolf Totem: A Novel*. Penguin.

Salz, Jeff: Adventure is Essential. https://www.jeffsalz.com, https://www.speakeradventure.com.

Singh Khalsa, Gurucharan, PhD. (2012). *The 21 Stages of Meditation*. Kundalini Research Institute.

St. Marie, Buffy. (2017). "Starwalker". Medicine Songs.

St. Marie, Buffy. (2017). "Spirit of the Wind". Medicine Songs.

Taimane. (2015). "Mother (Earth)". Elemental.

Thurston, Rik. Trance Channelling and Spiritual Teachings of the Transcendors, http://rikthurston.com.

Trungpa, Chogyam. (2009). *Smile at Fear: Awakening the True Heart of Bravery*. Shambhala.

Ushigua, Manari. Naku: https://www.naku.com.ec.

Wagamese, Richard. (2016). *Embers: One Ojibway's Meditations*. Douglas & McIntyre.

We Rise Together, The Indigenous Circle of Experts' Report and Recommendations, March 2018. Catalogue Number R62-548/2018E-PDF (ISBN 978-0-660-25571-2). http://www.publications.gc.ca/collections/collection_2018/pc/R62-548-2018-eng.pdf.

Whabagoon, Elder. https://whabagoon-flower-blooming-in-spring.business.site/?m=true.

Yin, Shin. (2014). *Quan Yin Speaks: Are you Ready?* XLIBRIS.

Zenato, Cristina. (2014, August 5). Cristina Zenato the 'Shark Whisperer'. https://youtu.be/G8LmxwOgBhA.

ABOUT THE AUTHOR

Sonia is a Canadian-Ukrainian lawyer, entrepreneur, and Heart-Centered Warrior who's spent more than fifteen years working in human rights, international law, business, economic development, community empowerment, as well as her own personal journey into herself.

Having spent the past seven years living and working with Indigenous nations around the world, as a facilitator, partner, shaman apprentice and friend, Sonia has gained a deep understanding of both ancient systems and modern ways, and our interconnection with all life.

Sonia is a certified kundalini yoga practitioner, energy healing facilitator, avid adventurer and explorer of the natural world. When not working, she spends her time at her home in the country where she grows her own food, forages, and interacts with the magic in nature to continually uncover its secrets.

www.soniamolodecky.com

www.globalindigenoustrust.org

BIOGRAPHIES

ELDER WHABAGOON—(FLOWER BLOOMING IN SPRING)

Elder Whabagoon is an Ojibway Elder and she sits with the Loon Clan. She is a member of the Lac Seul First Nation, a Keeper of Sacred Pipes, a 60's Scoop survivor, active community member, speaker, land defender, and water protector. Whabagoon hosts "Gatherings by the Fire" and "Sunrise Water Ceremonies" for those interested in sharing their love for the Land and our sacred Water and shares her teachings with Indigenous youth in a University of Toronto Access program she co-founded and co-leads called Nikibii Dawadinna Giigwag. Whabagoon has been recognized, awarded, and quoted for her strong commitment to educating the community and working with diverse groups, ages, and backgrounds. In her spare time, Whabagoon enjoys writing, painting, and spending time with her husband of 24 years, Karl, and their 5-year-old black cat, Theo.

ARZU MOUNTAIN SPIRIT

Ana Nicolasa Arzu, known as Arzu Mountain Spirit, is a traditional healer, writer, life coach, teacher, and presenter of talks and workshops on Garifuna spirituality, healing methods, and medicinal plants. She was born into a long succession of matriarchal Garifuna healers who will forever be her teachers. The Ancestral healing methods Arzu holds dear were transmitted through her mother, grandmother, and great grandmother. Arzu now lives in Belize where she maintains a traditional healing practice, and facilitates Mountain Spirit Wellness Retreats. In the effort to preserve and protect the treasures of her ancestors, she founded the Wagiya Foundation, a non-profit organization dedicated to the preservation of the Garifuna healing and spiritual traditions.